Modelling and Analysing the Market Integration of Flexible Demand and Storage Resources

Yujian Ye

东南大学出版社
·南京·

图书在版编目(CIP)数据

柔性负荷与储能资源接入电力市场的建模与分析技术 = Modelling and Analysing the Market Integration of Flexible Demand and Storage Resources：英文/叶宇剑著．—南京：东南大学出版社，2022.3
 ISBN 978-7-5641-9908-1

Ⅰ.①柔… Ⅱ.①叶… Ⅲ.①电力系统—英文 Ⅳ.①TM7

中国版本图书馆 CIP 数据核字(2021)第 263123 号

责任编辑：夏莉莉　　责任校对：张万莹　　封面设计：顾晓阳　　责任印制：周荣虎

Modelling and Analysing the Market Integration of Flexible Demand and Storage Resources

著　　者：叶宇剑
出版发行：东南大学出版社
社　　址：南京四牌楼 2 号　邮编：210096　电话：025 - 83793330
网　　址：http://www.seupress.com
电子邮件：press@seupress.com
经　　销：全国各地新华书店
印　　刷：江苏凤凰数码印务有限公司
开　　本：787 mm×1092 mm　　1/16
印　　张：12
字　　数：255 千字
版　　次：2022 年 3 月第 1 版
印　　次：2022 年 3 月第 1 次印刷
书　　号：ISBN 978-7-5641-9908-1
定　　价：68.00 元

本社图书若有印装质量问题，请直接与营销部联系。电话(传真)：025-83791830。

Foreword

The emerging Smart Grid paradigm has paved the way for the wide introduction of flexible demand (FD) and energy storage (ES) technologies in power systems, with significant economic, technical, and environmental benefits that will facilitate efficient transition to the low-carbon future. In the deregulated energy sector, the realization of the significant FD and ES flexibility potential should be coupled with their suitable integration in electricity markets. Previous studies have proposed several relevant market clearing mechanisms considering FD and ES participation and demonstrated their impacts on the system operation. However, these studies have neglected fundamental market complexities, such as modelling and pricing FD non-convexities as well as modelling and analysing the role of FD and ES in imperfect markets. This monograph is dedicated to addressing the above challenges through the development of novel computational methodologies.

Table of Contents

Chapter 1 Introduction ········ 1
  1.1 Monograph Context ········ 1
  1.2 Monograph Motivation ········ 3
      1.2.1 Role of Flexible Demand and Energy Storage in the Emerging Power System Setting ········ 3
      1.2.2 Market-Based Realisation of Flexible Demand and Energy Storage Flexibility ········ 4
  1.3 Monograph Scope and Original Contributions ········ 8
      1.3.1 Monograph Scope ········ 8
      1.3.2 Monograph Original Contributions ········ 8
  1.4 Monograph Outline ········ 13

Chapter 2 Current Trends on Market Integration of Flexible Demand and Energy Storage ········ 15
  2.1 Introduction ········ 15
  2.2 Modelling Approaches for Flexible Demand and Energy Storage ········ 15
  2.3 Market Arrangements for Flexible Demand and Energy Storage ········ 18
      2.3.1 Centralized Market Clearing Mechanism ········ 18
      2.3.2 Decentralized Dynamic Pricing-Based Mechanism ········ 19
  2.4 Participation of Flexible Demand and Energy Storage in Different Market Segments ········ 21
  2.5 Centralized Market Clearing Mechanism with Flexible Demand and Energy Storage Participation ········ 23
      2.5.1 Assumptions on Examined Electricity Market Model ········ 23
      2.5.2 Model of Centralized Market Clearing Mechanism ········ 28

Chapter 3 Factoring Flexible Demand Non-Convexities in Electricity Markets ········ 31
  3.1 Introduction ········ 31
  3.2 Literature Review ········ 32
  3.3 Centralized Market Clearing Under Flexible Demand Participation ········ 35

  3.3.1 Modelling Generation Participants 35
  3.3.2 Modelling Flexible Demand Participants 36
  3.3.3 Centralized Market Clearing Solutions 41
 3.4 Surplus Sub-Optimality Effects and Their Relation to Participants' Non-Convexities 42
  3.4.1 Surplus Sub-Optimality 42
  3.4.2 Generation Non-Convexities and Impacts on Surplus Optimality 44
  3.4.3 Flexible Demand Non-Convexities and Impacts on Surplus Optimality 45
 3.5 Generalized Uplifts Under Flexible Demand Participation 47
  3.5.1 Lump-Sum Uplifts 47
  3.5.2 Generalized Uplifts 48
  3.5.3 Formulation of Minimum Discrimination Problem 50
  3.5.4 Solution Techniques of the Minimum Discrimination Problem 53
 3.6 Convex Hull Pricing Under Flexible Demand Participation 55
  3.6.1 Concept of Convex Hull Pricing 55
  3.6.2 Lagrangian Formulation of Convex Hull Pricing Problem 57
  3.6.3 Solution Techniques of the Lagrangian Dual Problem 58
 3.7 Case Studies 62
  3.7.1 Test Data and Implementation 62
  3.7.2 Impacts of Flexible Demand Non-Convexities 64
  3.7.3 Generalized Uplift Approach 68
  3.7.4 Convex Hull Pricing Approach 69
 3.8 Conclusions 74

Chapter 4 Investigating the Impact of Flexible Demand and Energy Storage on the Exercise of Market Power by Strategic Producers in Imperfect Electricity Markets 78
 4.1 Introduction 78
 4.2 Literature Review 80
  4.2.1 Modelling Imperfect Markets with Strategic Electricity Producers 80
  4.2.2 Generation Market Power Mitigation 83
 4.3 Modelling Market Participants 85

  4.3.1 Strategic Generation Participants 85
  4.3.2 Demand Participants 88
  4.3.3 Energy Storage 91
 4.4 Theoretical Analysis of Impact of Demand Side and Energy Storage on Market Power 92
  4.4.1 Impact of Demand Own-Price Elasticity 92
  4.4.2 Impacts of Demand Shifting and Energy Storage 93
 4.5 Modelling Oligopolistic Electricity Markets with Demand Shifting and Energy Storage 95
  4.5.1 Bi-Level Optimization Model 96
  4.5.2 MPEC Formulation 99
  4.5.3 MILP Formulation 104
  4.5.4 Determining the Oligopolistic Market Equilibrium 107
 4.6 Case Studies 110
  4.6.1 Test Data and Implementation 110
  4.6.2 Impact of Demand Shifting and Energy Storage: Uncongested Network 112
  4.6.3 Impact of Demand Shifting and Energy Storage: Congested Network 117
 4.7 Conclusions 121

Chapter 5 Investigating the Exercise of Market Power by Strategic Flexible Demand and Energy Storage in Imperfect Electricity Markets 125
 5.1 Introduction 125
 5.2 Literature Review 127
  5.2.1 Modelling Strategic Behaviour of Demand Participants 127
  5.2.2 Modelling Strategic Behaviour of Energy Storage Participants 127
 5.3 Modelling Market Participants 130
  5.3.1 Generation Participants 130
  5.3.2 Demand Participants 130
  5.3.3 Strategic Energy Storage 131
 5.4 Qualitative Analysis of Demand Side and Energy Storage Market Power Capability 132
  5.4.1 Market Power Potential of the Demand Side 132
  5.4.2 Market Power Potential of Energy Storage 135

    5.5   Optimizing Capacity Withholding Strategies of Energy Storage ············ 136
           5.5.1   Bi-level Optimization Model ···································· 136
           5.5.2   MPEC Formulation ·· 139
           5.5.3   MILP Formulation ··· 140
    5.6   Case Studies ·· 142
           5.6.1   Test Data and Implementation ································ 142
           5.6.2   Quantifying the Optimal Extent of Capacity Withholding by Energy Storage ·· 145
           5.6.3   Impact of Storage Size ·· 147
           5.6.4   Impact of the Characteristics of the Demand Side ·············· 150
           5.6.5   Impact of the Characteristics of Wind Generation ············· 153
           5.6.6   Impact of Storage Location ······································ 154
    5.7   Conclusions ·· 156

Chapter 6   Conclusions and Future Work ··· 158
    6.1   Conclusions ·· 158
    6.2   Future work ·· 163
           6.2.1   Modelling and Pricing Flexible Demand Non-Convexities ······ 163
           6.2.2   Modelling and Analysing the Role of Flexible Demand and Energy Storage in Imperfect Markets ··· 165

References ·· 168

Appendix A   Convexity Principles ··· 177

Appendix B   Lagrangian Formulation of Convex Hull Pricing Problem ·········· 178

Appendix C   List of Abbreviations ·· 183

Chapter 1
Introduction

1.1 Monograph Context

Contemporarily, electrical power systems across the world are facing unprecedented challenges. First of all, climate change resulting from greenhouse gases (GHG) emissions poses a huge threat to human welfare. To contain such threat, governments around the world have committed themselves to substantial reductions in GHG emissions in the coming decades. In the United Kingdom (UK), aggressive long-term climate change targets have been incorporated in the current government energy policies. The *2008 Climate Change Act* defined a legally binding target to reduce GHG emissions by 34% by 2020 and by at least 80% by 2050, compared to 1990 baseline levels[1].

Secondly, energy security also has to be considered alongside climate change, with the aim of trying to ensure that both of these issues are collectively addressed. The growing energy security concern arises over the reliance of electricity generation on fossil fuels exhibiting a continuously declining availability and a subsequent price growth. As part of the efforts to tackle the above climate and energy security concerns, energy systems are undergoing a transition towards deep decarbonisation, by virtue of the wide integration of intermittent renewable generation (such as wind and solar) conjointly with low-carbon generation (such as nuclear and Carbon Capture and Storage (CCS) enabled generation). The European Commission (EC) has issued a directive in 2009 which put forward legally binding targets on the share of renewable energy source (RES) in the overall energy supply for individual member states of the European Union (EU)[2]. The goal of the directive at the EU level is for RES to cover 20% of the total energy demand in 2020. The UK has set a target to produce at least 15% of its energy consumption from RES by 2020, representing a marked increase from the 2014 level of 7.1% and 2015 level of 8.3%[3]. However, the majority of RES are characterized by inherent variability, unpredictability and non-controllability. In the UK, wind generation (both on- and offshore), presently constitutes the most commercially mature and scalable renewable technology with a significant projected growth through to 2020 and beyond[4]. Furthermore, as suggested by the *2016 UK Future Energy Scenarios*, to meet the 2050 decarbonisation target, emissions need to reduce from 2015 levels by 70% in the transport sector, and 65% in the heat sector respectively[4]. To support the ongoing and future decarbonisation of electricity generation systems, strong motives arise for the *electrification* of large segments of transport and heat demand, and source whose supply from low-carbon and RES generation. Regarding the former, recent advancements made in automotive and battery technologies have paved the way for the introduction of *electric vehicles* (EV)[5-6], reducing usage of fossil fuels (and thus reducing carbon emissions) of the traditional internal combustion engine vehicles. Concerning the latter, recent technological developments in advanced heating systems have enabled the transition from traditional gas/oil-fired systems to the promising and energy efficient systems employing *electric heat pump* (EHP) technology[7]. However, due to the energy-intensive features of transportation vehicles and heating loads, the climate and energy security potential of this sectoral shift is accompanied by a potential increase in peak demand that is disproportionately higher than the increase in total energy consumption. Given the significant penetration of low capacity value RES generation, accompanied with the introduction of such new demand, the future electricity system could be characterized by considerable increased generation and network costs and degraded asset utilization[8]. Moreover, meeting the future electricity demand is likely to rely on the use of non-renewable and low-carbon generation technologies such as nuclear and CCS plants, which have lower operational flexibility compared to the existing coal and combined-cycle gas turbine (CCGT) plants[9-10].

Thirdly, the decarbonisation of both energy generation and demand sectors introduces significant challenges to the operation and development of future power systems. In line with the effort to address the challenges of decarbonisation through the wide integration of RES generation, greater need for flexibility stems. Traditionally, operational flexibility has been mainly provided by conventional generators that were flexible enough to follow variations in demand and adjust its output in response to contingencies. Electricity produced from intermittent RES however is highly variable in time, difficult to predict and cannot be controlled apart from curtailing its electricity output. This is very different from the majority of conventional generation that not only provides energy but also vital control services (e.g. balancing and network support services) necessary to maintain integrity of the power system. Furthermore, given that renewables represent non-synchronous power sources that do not contribute to system inertia, the overall system inertia will decrease which will increase requirements for primary frequency regulation. Overall, unpredictability and lack of inertia associated with RES will impose a very considerable requirement for additional flexibility, particularly ancillary services associated with real time balancing. Increased requirements for real time ancillary services, if provided by conventional generation running part-loaded, will not only reduce the efficiency of system operation but also may significantly undermine the ability of the system to absorb intermittent RES output.

Going further, the *Electricity Act 1989* laid the legislative foundations for the restructuring and privatisation of the electricity industry in Great Britain (GB)[11-12]. Since then, significant efforts towards the deregulation of the electricity industry have been made over the last two decades, characterized by the unbundling of vertically integrated utilities and promoting competition in the generation, supply and network sectors of the industry[11]. In this context, suitably designed competitive electricity markets are expected to facilitate efficient operation and development of power systems including integration of significant amounts of RES.

Finally, numerous advances have been made in many technology sectors in recent years. Among others, these include efficient generation, demand and energy storage technologies, power electronics, advanced metering, innovative control systems as well as modern information and communication technologies (ICT). The application of these technologies promises considerable benefits for a more economically efficient operation and development of power systems, but at the same time presents significant challenges, particularly in terms of designing suitable financial and regulatory incentives for their large-scale deployment.

1.2 Monograph Motivation

1.2.1 Role of Flexible Demand and Energy Storage in the Emerging Power System Setting

In the emerging power system setting described in the previous section, it is manifest that meeting the future needs for flexibility with solutions solely based on conventional network and generation capacity might become very inefficient and costly while also potentially hindering the delivery of carbon reduction targets. Ever more research efforts consequently are focused on the alternative sources of flexibility, such as emerging designs of *flexible generation technologies*, *flexible network technologies*, *flexible demand* and *energy storage technologies*[10]. This monograph focuses on the last two as their significant flexibility potential improves the resource-efficiency of the electricity system, and thus contribute greatly to the realization of the vision for the future power systems.

Flexible Demand (FD)[10], [13-14], e.g. EVs, EHPs, Wet Appliances (WAs), typically involves temporal redistribution of consumers' energy requirements. As a large number of researchers have stressed, consumers' flexibility regarding electricity use mainly involves shifting of their loads' operation in time instead of simply avoiding using their loads[15-16]. In other words, load reduction during certain periods is accompanied by a *load recovery effect* during preceding or succeeding periods. This shift of energy demand from different periods drives a demand profile flattening effect.

Energy Storage (ES)[9], [17], typically involves the conversion of electricity into another form of energy suitable for storing (e.g. potential, chemical, kinetic, compressed air etc.) at one time period and for use at a later time period. The charging and discharging operation of ES drives a similar shift of energy demand across time, enables flattening of demand profile (as with the FD operation), and thus exhibits similar time-shifting flexibility potential.

If such flexibility potential is suitably exploited, it can induce significant economic, technical and environmental benefits for the future power systems. These include the support of system balancing and thus avoid inefficiency of part-loaded conventional generation and/or curtailment of wind generation, the reduction of demand peaks from electrification of transport and heat sectors and thus reducing generation operational costs, postponing or avoiding capital-intensive generation and network capacity investments[8-10], [13-19].

1.2.2 Market-Based Realisation of Flexible Demand and Energy Storage Flexibility

The deployment of the demand flexibility potential is not a new concept. Before the deregulation of the electricity industry, strategies for the integration of FD in system operation and development were adopted. Under such schemes, the regulated utility would manage and optimise this flexibility in the form of centrally administrated programs[20]. In the post-deregulation era however, the operation and development of the power systems relies on appropriately designed competitive electricity markets. In this environment, the realization of FD and ES flexibility potential should be coupled with their integration in electricity markets. Driven by competitive market dynamics, suitable integration schemes should take into account the effect that FD and ES may have in the market operation, particularly, their contributions to the market price-setting process. Such schemes would enable demand side and storage to actively participate in the market, which would make electricity markets more efficient and more competitive, and would also support a more optimal allocation of economic resources[21].

Approaches examined in the literature that achieve integration of FD[22-27] and ES[28-32] in electricity markets generally extend the traditional centralized mechanism to enable a two-side active participation. In this setting, FD and ES are treated as active market participants that are capable of making rational decisions as generation participants. A central entity, indicated as the market operator, collects bids/offer from generation, FD and ES participants that comprise information regarding their individual technical and economic characteristics. Using the collected knowledge, the market operator clears the market by performing a global optimization (typically with the objective of social welfare maximization) that yields the optimal market outcome from the system perspective. While actively participating in the market, the operating decisions for FD and ES are established on an overall efficiency target to maximize social welfare, consequently this scheme enables to clearly determine the potential benefits in social welfare obtained by integrating FD and ES into the system operation.

The above centralized mechanism achieves market-based realization and exploitation of FD and ES flexibility potential. In this context, references [9], [13], [19], [22]—[32] have employed such mechanism and demonstrated multiple value streams of FD and ES for the future power systems. Among others, this includes the reduction in generation operating costs, the reduction of the generation capacity margin, the enhancement of electricity supply reliability, the reduction and deferral of investment in both generation and network assets, the improvement of distributed network investment efficiency and congestion management, the support of system balancing and the provision of ancillary services with high penetration of intermittent RES, etc. Although the previously mentioned studies have set a basic framework for modelling and analysing the participation of FD and ES in electricity markets, it has neglected fundamental market complexities, such as modelling and pricing FD non-convexities as well as modelling and analysing the role of FD and ES in imperfect markets. These shortcomings of existing work are further discussed below:

Modelling and pricing FD non-convexities

In the emerged deregulated market environment, a concern arises as to whether the market arrangements will provide appropriate incentives to the self-interested market participants to deliver the social welfare maximizing system operation solutions. As various economic literature has demonstrated, in markets characterized by *non-convexities*, the traditional *marginal pricing* mechanism cannot generally support *competitive equilibrium* solutions[33-45]. In other words, individual market participants' surplus maximizing self-schedule given the prices determined by the centralized market clearing problem, is not generally consistent with the schedule calculated by the latter. In cases of such inconsistencies, the centralized schedules entail lower surpluses than self-scheduling, with this difference termed as *surplus loss* and the related effect as *surplus sub-optimality*. A wide literature has identified non-convexities associated with the generation side of electricity markets, including binary (on/off) commitment decisions, fixed and start-up/shut-down costs, minimum stable generation constraints, and minimum up/down times constraints[34-45]. The relation of these non-convexities with inconsistency and surplus sub-optimality effects has been demonstrated. Different approaches have been explored to address these effects. Among others, these include lump-sum uplifts, convex hull pricing, and non-linear and differentiated pricing mechanisms[34-45].

Despite the recent research and industry interest in FD technologies due to their significant economic and environmental potential, the demand side has not been explored from the same perspective. Even simple FD models reveal non-convexities in the operation characteristics of flexible loads, such as the option to forgo demand activities[15], discrete operating characteristics of wet appliances[14], etc. Therefore, strong motives arise for systematically identification of these FD non-convexities and demonstration of their relation with potential FD inconsistency and surplus sub-optimality effects. It also necessitates appropriate modification in previously developed pricing mechanisms for addressing non-convexities of the generation side, to account for FD non-convexities.

Modelling and analysing the role of FD and ES in imperfect markets

After the deregulation of the energy sector, electricity markets have been gradually moved to a more liberalized structure. In this context, the restructured electricity markets are better described in terms of imperfect rather than perfect competition. In this setting, market participants do not necessarily act as *price takers*[46]. Participants owning a large share of the market or strategically located in the electricity network are able to affect and manipulate the electricity prices and increase their surplus beyond the competitive equilibrium levels, through strategic bids and offers. This exercise of market power brings market performance far from the perfect competition equilibrium, and results in loss of social welfare[46]. Market power mitigation is therefore among the major concerns in electricity market design and operation.

Previous work on imperfect electricity markets has mainly focused on the generation side, by developing methodologies for optimizing bidding strategies of individual electricity producers as well as investigating market equilibria resulting from the interaction of multiple strategic producers[47-55]. Moreover, various remedies to mitigate strategic gaming behaviour of the generation side have been suggested. Among others, encouraging demand response is regarded as a promising way towards more competitive electricity markets. The role of the demand side in this context has been previously investigated in terms of the effect of its *own-price elasticity*[25] on electricity producers' ability to exercise market power. However, the concept of own-price elasticity cannot fully capture consumers' flexibility regarding energy use, as they are more likely to shift their loads' operation in time instead of simply avoiding it[15], [21]. This *time-shifting flexibility* will be enhanced with the penetration of various FD and ES technologies, envisaged by the *Smart Grid* paradigm[56]. Although numerous recent studies have investigated the impact of this flexibility on assorted aspects of power system operation and planning, its role in imperfect electricity markets has not been explored yet.

Furthermore, the ability of FD and ES to exercise market power needs to be better analysed. Regarding the former, previous work has demonstrated that large consumers can strategically increase their surplus by submitting bids lower than their actual marginal benefit[57-58], while other market power potentials of FDs are not explored. Concerning the latter, previous work [59]—[62] has demonstrated the ability of large ES units to exercise market power by withholding their capacity, leading to additional ES profits but loss of social welfare. However, the modelling approaches adopted in references [59]—[61] exhibit certain limitations, and the dependency of the extent of exercised market power (i.e. the additional ES profit and loss of social welfare) on ES and system parameters are not properly analysed in previous studies.

1.3 Monograph Scope and Original Contributions

1.3.1 Monograph Scope

This monograph focuses on modelling and pricing FD non-convexities as well as modelling and analysing the role of FD and ES in imperfect markets.

Modelling and pricing FD non-convexities

This research lies in systematically identifying non-convexities associated with the operation of FD, and analysing the relation of the former with schedule inconsistency and surplus sub-optimality effects. Previously proposed pricing mechanisms addressing non-convexities of the generation side will be suitably extended to account for FD non-convexities.

Modelling and analysing the role of FD and ES in imperfect markets

The first aspect of this research lies in analysing the impact of FD and ES on the extent of market power exercised by strategic producers. The second aspect of this research lies in exploring the ability of FD and ES to exercise market power.

The qualitative and quantitative insights provided in this monograph are crucial for the envisaged enhancement of the role of FD and ES in the emerged deregulated power system environment.

1.3.2 Monograph Original Contributions

In the context of the scope discussed in subsection 1.3.1, the original contributions of this monograph are associated with the development, analysis and testing of suitable models, methods, and examples to deal with the challenges associated with modelling and pricing FD non-convexities as well as modelling and analysing the role of FD and ES in imperfect markets, as described in subsection 1.2.2. These original contributions are further explored as follows:

Modelling and pricing FD non-convexities

- Detailed operational models are derived to represent two different types of FDs, capturing the largest part of flexible load characteristics in the related literature: *continuously-controllable* FD (CCFD) and *fixed-cycle* FD (FCFD). The power demand of a CCFD can be continuously adjusted between a minimum and a maximum limit when the FD is active. FCFDs involve operating cycles which comprise a sequence of phases occurring at a fixed order, with fixed duration and fixed power consumption, which cannot be altered.

- Two different demand flexibility potentials are modelled. The first is associated with the ability to completely forgo demand activities and the second is associated with the ability to redistribute the total electrical energy requirements of activities across time. All FDs are assumed to exhibit both flexibility potentials.

- Based on a thorough examination of the characteristics of the FD operational models, non-convexities of FD are identified, including options to forgo demand activities, minimum power levels associated with CCFDs as well as discrete power levels associated with FCFDs. The FD non-convexities are different to the respective non-convexities associated with the generators given their significant operational differences.

- The relation of these FD non-convexities with schedule inconsistency and surplus sub-optimality effects is demonstrated through simple one- and two-time period examples and a larger case study with day-ahead horizon and hourly resolution. The analysis reveals that in the presence of FD non-convexities and under the traditional marginal pricing mechanism, the FDs concentrate at the lowest-priced hours within their scheduling period under self-scheduling, which is not consistent with the centralized schedule involving an as-flat-as-possible total demand profile. The above inconsistencies are then translated into FD surplus losses.

- Generalized uplift and convex hull pricing approaches addressing these schedule inconsistency and surplus sub-optimality effects are suitably extended to account for FD non-convexities. Concerning the former, generalized uplift functions for FD participants are proposed, which include FD-specific terms and constitute additional benefits or payments for the FDs. The structure of the generalized uplift function of FDs is not the same with the respective function for generators, due to their different non-convexities and resulting surplus sub-optimality effects. The parameters of these functions along with the electricity prices are adjusted to achieve consistency for every FD participant. A new rule is introduced for equitable distribution of the total generators' profit loss and FDs' utility loss compensation among market participants. Regarding the latter, convex hull prices are calculated as the Lagrangian multipliers optimizing the dual problem of the market clearing problem considering FD participation. It is demonstrated that convex hull prices are flattened at periods when FD is scheduled to eliminate surplus sub-optimality associated with the FD ability to redistribute energy requirements across time.

Modelling and analysing the role of FD and ES in imperfect markets

- The first contribution of this research lies in providing for the first time both theoretical and quantitative evidence of the beneficial impact of FD and ES in limiting market power exercised by strategic electricity producers.

- Theoretical explanation of this impact is presented through a price-quantity graph on a simplified two-period market. It is demonstrated that the operation of FD and ES drives a demand profile flattening effect by reducing the peak demand while increasing the off-peak demand. This in turn reduces the price increment (driven by the exercise of market power of strategic producers) at the peak period while increases it at the off-peak period. However, the price increment reduction at the peak period is more prominent than its increase at the off-peak period, due to the larger slope of the marginal cost curve of generation. This effect implies that the deployment of FD and ES results in an overall reduction of strategic producers' ability to manipulate market prices.

- Quantitative analysis is supported by a multi-period *equilibrium programming* model of the oligopolistic market setting. The decision-making process of each strategic producer is modelled through a bi-level optimization problem. The upper level (UL) represents the profit maximization problem of the producer and the lower level (LL) represents endogenously the market clearing process, accounting for the time-coupling operational constraints of FD and ES, and network constraints. This bi-level problem is converted to a *Mathematical Program with Equilibrium Constraints* (MPEC), by replacing the lower level problem by its equivalent *Karush-Kuhn-Tucker* (KKT) optimality conditions. By employing suitable linearization techniques, this non-linear MPEC is in turn converted to a *Mixed-Integer Linear Program* (MILP) which can be solved using available commercial solvers. An *iterative diagonalization* approach is employed in order to determine market equilibria resulting from the interaction of multiple independent strategic producers.

- Case studies with the developed model on a test market reflecting the general generation and demand characteristics of the GB system have quantitatively demonstrated the benefits of FD and ES in limiting producers' market power, by employing relevant indexes from the literature. In cases without network congestion, the location of FD and ES flexibility does not have an impact on producers' market power exercise, but an increasing FD flexibility and size of ES is shown to i) reduce strategic producers' ability to manipulate market prices, ii) reduce strategic producers' additional profit driven by the exercise of market power, iii) allow consumers to more efficiently preserve their economic surplus against producers' strategic behavior, and iv) reduce the social welfare loss and thus enhance the overall efficiency of the market. In cases with network congestion, FD and ES flexibility still has an overall positive impact on market efficiency, but the extent of this benefit as well as the impact on producers and demands at different areas depends on the location of FD and ES in the network.

- The second contribution of this research lies in exploring the market power potential of FD and ES.

- Regarding FD, its ability to exercise market power by revealing less time-shifting flexibility to the market is qualitatively analysed through a price-quantity graph in a simplified two-period market. It is shown that the competitive demand shifting behaviour enables a net demand profile flattening effect by reducing/increasing demand at the peak/off-peak period respectively. By acting strategically and reporting less time-shifting flexibility to the market, the demand participant limits this flattening effect on the system demand profile, since it increases/reduces less the demand at the off-peak/peak period respectively. This strategic action has a more dominant negative impact on demand payment at the peak period (due to the larger slope of the marginal cost curve of generation) than its positive impact at the off-peak period, resulting in an overall higher demand payment. As a result, consumers will find it financially unattractive to behave in such a strategic fashion.

- Concerning ES, this monograph develops a multi-period game-theoretic model for optimizing capacity withholding strategies of ES. The decision-making process of strategic ES is modelled through a bi-level optimization problem. The upper level represents the profit maximization objective of the strategic ES and the lower level represents endogenously the market clearing process, accounting for the network constraints. This bi-level problem is converted to a MPEC, by replacing the lower level problem by its equivalent KKT optimality conditions. Suitable linearization technique is proposed to convert this non-linear MPEC to a MILP which can be efficiently solved using available commercial solvers.

- Theoretical analysis in a simplified two-period market as well as case studies with the developed optimization model on a test market with day-ahead horizon and hourly resolution demonstrate that ES needs to optimize the extent of exercised capacity withholding, in order to achieve the best trade-off between the positive effect of capacity withholding on the peak to off-peak price differential and its negative effect on the volume of energy sold by ES. Due to this negative effect, the optimal extent of capacity withholding is different at different periods of the market horizon, being lower during periods when ES charges and discharges higher energy.

- Case studies have adopted relevant indexes to measure the extent of market power exercised by the energy storage. The impact of the size of energy storage (in terms of its power rating and energy capacity) on the extent of exercised capacity withholding and the resulting storage profit increase and social welfare loss is analyzed. On the one hand, a higher power rating increases the extent of exercised capacity withholding as ES attempts to maintain the peak to off-peak price differential at a high level. On the other hand, a higher energy capacity reduces the extent of capacity withholding as ES is motivated to act more truthfully in order to exploit its higher energy content and sell more energy. Nevertheless, both a higher power rating and a higher energy capacity increase the additional profit made by ES and the social welfare loss (with respect to the case where ES acts competitively), since they increase the ability of ES to affect market prices. This trend however exhibits a saturation effect, as increase of the power rating and the energy capacity beyond a certain level does not affect the market outcome under neither competitive nor strategic ES behaviour.

- Case studies also analyse the impact of the generation and demand characteristics of the market on the extent of market power exercised by ES. Concerning the demand side, a higher price elasticity and a flatter demand profile reduce the additional profit made by ES and the social welfare loss. This result indicates that the envisaged realization of demand response measures enhancing the consumers' price responsiveness and flattening the demand profile by rescheduling consumers' demand towards off-peak periods will have a positive effect in limiting the extent of market power exercised by ES. Regarding the generation side, a higher penetration of wind generation suppresses price levels and therefore generally mitigates the extent of market power exercised by strategic ES.

- Finally, the studies demonstrate that the location of ES affects its ability to exercise market power if the network over which the market operates is congested. In this case, the market power potential of ES is significantly increased when it is located in areas with more expensive generation and higher demand.

1.4 Monograph Outline

The remainder of this monograph is organized as follows:

Chapter 2 first discusses FD and ES operational models in electricity markets, and details some of the most significant assumptions employed in the current modelling approaches. Next, it outlines the state-of-the-art market clearing mechanisms achieving integration of FD and ES in electricity markets. It subsequently reviews relevant literature analysing simultaneous participation of FD and ES in multiple market segments to provide multiple system services. Finally, this chapter discusses the key assumptions made regarding the design and participants' behaviour of the examined electricity markets, and presents a fundamental formulation of the centralized market clearing mechanism adopted throughout the monograph for achieving integration of FD and ES.

Chapter 3 in the first place conducts a comprehensive literature review on previously proposed pricing mechanisms in electricity markets with generation non-convexities. It then derives detailed operational models of FD and formulates the centralized market clearing problem under FD participation. Next, it systematically identifies FD non-convexities and demonstrates the resulting schedule inconsistency and surplus sub-optimality effects through simple examples. It subsequently details the extension of generalized uplift and convex hull pricing approaches respectively to account for both generation and FD participation. Finally, this chapter presents the examined case study, where participants' schedule inconsistency and surplus sub-optimality effects in the presence of non-convexities are illustrated, and the effectiveness of generalized uplift and convex hull pricing approaches to deal with these effects are demonstrated.

Chapter 4 first presents a comprehensive literature review regarding modelling approaches of imperfect markets with strategic electricity producers and previously proposed market power mitigation approaches. It then outlines operational models of strategic generation, demand and storage market participants. Next, it provides a theoretical explanation of the beneficial impact of FD and ES on market power through a price-quantity graph in a simplified two-period market. It then details the bi-level optimization model expressing the decision-making process of each strategic electricity producer. The analytical derivation of the equivalent MPEC and MILP formulations is subsequently presented. Going further, it introduces the iterative diagonalization approach employed to identify the oligopolistic market equilibria resulting from the interaction of multiple independent producers. Finally, case studies on a test market are presented, quantitatively demonstrate the benefits of FD and ES in limiting the exercise of market power for different scenarios regarding: i) the time-shifting flexibility of FD, ii) the size of ES, iii) the location of FD and ES in the network, and iv) network congestion, by employing relevant market power metrics from the literature.

Chapter 5 first conducts a detailed literature review regarding modelling strategic behaviour of the FD and ES in deregulated electricity markets. It then outlines operational models of generation, strategic demand and storage market participants. Next, it qualitatively analyses the market power potential of the FD and ES through a price-quantity graph in a simplified two-period market. It then formulates the bi-level optimization model expressing the decision-making of strategic ES and explains the derivation of the equivalent MPEC and MILP formulations. Finally, case studies on a test market are presented, quantitatively analyse the extent of capacity withholding and its impacts on ES profit and social welfare at different time periods and for different scenarios regarding: i) the size of ES (in terms of its power rating and energy capacity), ii) the characteristics of the demand side (in terms of its profile shape and price elasticity), iii) the characteristics of the generation side (in terms of the installed capacity and output profile shape of wind generation), and iv) the location of energy storage in the presence of network congestion, resorting to relevant market power indexes from the literature.

Chapter 6 concludes this monograph by reviewing the main contributions of this research and identifying interesting areas for future research.

Chapter 2
Current Trends on Market Integration of Flexible Demand and Energy Storage

2.1 Introduction

The emerging Smart Grid paradigm[56] has paved the way for the wide introduction of FD and ES technologies in power systems, with significant economic, technical, and environmental benefits that will facilitate efficient transition to the low-carbon future. In the deregulated energy sector, the realization of the significant FD and ES flexibility potential should be coupled with their suitable integration in electricity markets. In this context, this chapter discusses a number of important topics regarding current trends on market integration of FD and ES.

This chapter is organized as follows. Section 2.2 reviews FD and ES operational models in electricity markets, and details some of the most significant assumptions employed in the current modelling approaches. Section 2.3 outlines state-of-the-art in market-based FD and ES integration including centralized market clearing mechanism and decentralized dynamic pricing-based mechanism. Section 2.4 reviews relevant literature analysing simultaneous participation of FD and ES in multiple market segments to provide multiple system services. Finally, Section 2.5 lays out the main assumptions made regarding the design and participants' behaviour of the examined electricity markets, and presents a fundamental formulation of the centralized market clearing mechanism adopted throughout the monograph for achieving integration of FD and ES.

2.2 Modelling Approaches for Flexible Demand and Energy Storage

As discussed in Chapter 1, suitable market-based realisation of FD/ES flexibility potential could promise significant benefits for the future power systems. However, a lack of understanding and experience with these new participants has necessitated the employment of assumptions in the modelling approaches adopted. As a result, the quantification of impacts and benefits of these participants will be dependent on these assumptions, and an accurate evaluation has yet to be achieved[18]. This section presents a general literature review concerning FD and ES models in electricity markets, and details some of the most significant assumptions used.

In reference [57], FD is modelled similarly to negative generation with pick-up and drop ramping limits, minimum and maximum consumption constraints, as well as minimum total energy requirements. In reference [63], the demand side is classified into load clipping and load shifting categories, which represents two general behaviours of demand response: curtailment and deferral of demand. The *pay-back effect* of the demand side is modelled in reference [64], where a reduction of the demand at a specific time period is followed by an increase of the demand at a subsequent time period. References [14], [66]—[68] look at technology specific FDs including EVs with flexible charging capability, space-heating EHPs incorporating heat storage, and WAs with deferrable initiation time. In reference [59], ES is considered as a part of the portfolio of a generation company, while reference [60] examines different ownerships of ES including generators, consumers, or merchant storage operators. In reference [68] and reference [69], ES is modelled as a part of a storage-solar and storage-wind coalition respectively. On the other hand, references [61]—[62], references [70]—[71] treat ES as an independent market participant. The operational characteristics of ES generally include energy capacity limits, energy content (or state of charge (SOC)) balance accounting for charging/discharging losses, and maximum charging/discharging rates[28-32], [59-62], [70-71].

Aside from the above predominated approaches, some studies assume that the demand side behaves in a purely economically rational manner. Based on the concept from microeconomics[72], this behaviour can be expressed by a linear decreasing demand function that captures the effect of *own-price elasticity* (which is defined as the ratio of the relative change in demand to the relative change in price). Given this own-price elasticity value, an aggregated demand-bid curve can be constructed which allows the (price) responsiveness of demand to be incorporated into the market clearing model. However, modelling demand response based on single-period own-price elasticity suggests that consumers can only increase or decrease their loads, while the operation of their loads cannot be shifted across time. In order to capture such inter-temporal behaviour, references [25], [73]—[74] propose the use of *cross-price elasticity*, which considers the shift of demand to another time period due to a change in price at the current period. In other words, modelling both self- and cross-price elasticity takes into account the fact that energy which is not consumed now, through a reduction of loads, must be recovered later. However, reference [73] argues that the consideration of the cross-price elasticity may reduce the attainable demand response to some extent. The authors examined a case where several adjacent time periods have similarly high prices. In this case, the demand curtailed in one period is shifted to another period, or over multiple periods, and this occurs for each of the periods during which the price is high. This leads to the combined effect that some demand at a specific period is reduced, but demand from many other periods may be shifted to this period. As a consequence, the total demand response attainable is reduced in comparison to the case where only the own-price elastic behaviour is exhibited. This work only stands as an illustration of the impact of considering different forms of price elasticity. However, as demand response is not well represented in the form of a self- and cross-price elasticity matrix, more detailed modelling is required to achieve a realistic representation of demand side flexibilities.

The modelling approaches developed in this monograph constitute an attempt to contribute to a more in-depth understanding of the actual flexibility potential of FD and ES. Chapter 3 develops detailed operational models, in a non-technology specific way, to represent two different types of FDs, capturing the largest part of flexible load characteristics in the related literature: continuously-controllable FD (CCFD) and fixed-cycle FD (FCFD). Both types are flexible in terms of the time period(s) that they can obtain the amount of energy required for their operation, as long as this is done within a temporal interval allowed by their users. For the first type, the power demand level can be continuously adjusted between a minimum and a maximum limit when the FD is active. The operation of the second type is based on the execution of user-called cycles which comprise a sequence of phases occurring at a fixed order with generally fixed duration and fixed power demand that cannot be modified; their flexibility involves the deferability of these cycles up to a maximum delay limit set by their users. In Chapter 4 and 5, we model the marginal benefit or the willingness to pay as a linear decreasing function which captures the effect of demand side's own-price elasticity. The time-shifting flexibility of demand side is also captured in the proposed operational models, where the change of demand at each time period due to load shifting is expressed as a ratio of the baseline demand. While demand shifting is energy neutral within the examined time horizon i. e. the load reduction during certain periods is always accompanied by a load recovery during preceding or succeeding periods. While the energy payment depends on the final (net) demand after any potential load shifting, the benefit obtained by the demand side (i.e. the level of service delivered to the consumers) is assumed to depend on only the baseline demand level; this assumption expresses the flexibility of the consumers to shift the operation of some of their loads without compromising the satisfaction they experience. Our model captures more accurately the actual flexibility potential of electrical consumers comparatively to the cross-price elasticity model. Inline with references [61]—[62], [70]—[71], ES is modelled as an independent market participant and of significant size, so that its operation affects market clearing outcomes (e.g. market prices), and consequently affects the strategic behaviour of electricity producers (Chapter 4) as well as its own profit (Chapter 5).

2.3 Market Arrangements for Flexible Demand and Energy Storage

Market-based approaches examined in the literature that achieve integration of the FD and ES in electricity markets can be broadly classified into two categories: *centralized* market clearing mechanism and *decentralized* dynamic pricing based mechanism.

2.3.1 Centralized Market Clearing Mechanism

As discussed in subsection 1.2.2, the first approach revolves around the extension of traditional centrally administered mechanisms by enabling a two-side active participation. Generation, FD/ES participants submit their economic and technical characteristics in the form of bids and offers to the market operator. The latter clears the market through the global solution of a social welfare maximization problem that yields the optimal market outcomes from the system perspective. In this context, FD participants may generally correspond to large industrial/commercial consumers, participating individually in the market clearing, or FD aggregators, representing a large number of smaller residential consumers[75-76]. Similar aggregation of a number of small-scale distributed ESs is presented in reference [77], whereas references [29], [61]—[62], [70]—[71] consider the participation of large-scale bulk ESs in the market.

The operating decisions for FD and ES under such centralized mechanism are established on an overall efficiency target to maximize social welfare, therefore this scheme enables to clearly determine potential benefits in social welfare obtained by integrating FD and

ES into the system operation. Previous work (references [22]—[24], [28]—[29]) has adopted such centralized mechanisms considering FD and ES participation and demonstrates their impacts on assorted aspects of system operation. In reference [22], the hourly demand response with inter-temporal characteristics is incorporated into a security-constrained unit commitment (SCUC) model. The case studies demonstrate that demand response contributes to peak load shaving, the reduction of the system operating cost, and the reduction of transmission congestion through reshaping the hourly demand profile. A smart grid model is presented in reference [23], it is demonstrated that plug-in electric vehicles (PEVs) contribute to maximize utilization of RESs to reduce costs and emissions from the electricity industry. In reference [24], an economic dispatch (ED) problem is formulated considering price responsive and demand shifting bids. The developed bidding mechanisms contribute to improve congestion management. Authors in reference [28] proposed a stochastic unit commitment model with bulk energy storage, and demonstrated the ability of the latter to reduce the total generation cost for a system with high renewable penetration level. In reference [29], the optimal capacity and operation for a large-scale storage system is optimized in a centrally controlled market in order to minimize the total system operation cost. The results demonstrate that the operation of ES reduces the market price volatility and supports the integration of RES generation.

The main weakness of the centralized mechanisms is the lack of communication and computational scalability[14], [65]. For instance, under significant FD participation, transmission of complex and diverse operating parameters and constraints of a very large number of flexible loads to the central market operator will result in information collection and communication challenges, while the vast number of decision variables and constraints associated with such loads in the optimization problem will impose considerable computational burden to the market operator. Furthermore, centralized mechanisms could raise privacy objections by the consumers, as they necessitates the revelation of sensitive information (e.g. preferences, habits, loads' characteristics) of the electrical consumers to the central market operator.

2.3.2 Decentralized Dynamic Pricing-Based Mechanism

To resolve the limitations associated with the centralized mechanisms, the second approach involves decentralized market participation of FD and ES. The decentralized mechanisms generally incorporate *dynamic pricing* schemes[20], [78-81], that expose FD and

ES to pre-determined, time-varying prices reflecting more accurately the actual marginal cost of supply. The objective of such mechanism is to encourage FD and ES participation based on price differentials across different time periods so as to promote more efficient markets. Dynamic pricing schemes general include *Time of Use* (TOU) rate, *Critical Peak Pricing* (CPP), and *Real Time Pricing* (RTP)[20]. The TOU rates vary by time of the day. The rate during peak periods is higher than the rate during other off-peak periods. The design of TOU rate is aiming towards reflecting the average cost of electricity during different periods. CPP rates generally include a pre-specified higher electricity usage price superimposed on TOU rates, which are called during system contingencies. RTP rates typically fluctuate hourly, reflecting changes in the wholesale price of electricity.

Under such time-differentiated pricing schemes, the market integration of FD and ES is induced voluntarily (i.e. participants do not report their individual characteristics to a central entity, and act as price-takers in the market), as opposed to the centralized schemes. On the basis of the posted price information, FD and ES are encouraged to exploit their flexibility potential so as to optimize their operational decisions according to their individual interests (i.e. maximization of their economic surpluses), as opposite to the case under the centralized mechanism where the operational decisions of FD and ES are established from the system perspective, while their profitability is not really addressed. The communicated prices are higher during peak demand periods and lower during off-peak demand periods to motivate FDs to reschedule their demand towards the latter (ES is incentivized to charge during the former and discharge during the latter, driving similar temporal shift of energy demand), and therefore improve the efficiency of system operation. Relevant studies [78]—[81] that achieve decentralized participation of FD and ES with price-based control approaches are outlined as follows. A robust optimization model is developed in reference [78] to adjust the hourly load level of a given consumer in response to hourly electricity prices. The objective of this model is to maximize the utility of the consumer subject to the operational constraints of its loads. Authors in reference [79] developed a dynamic tariff with the purpose of alleviating distribution network congestion resulting from simultaneous EV charging. In accordance to the published locational prices, EVs optimize their charging schedules in order to minimize their charging costs. The optimized EV charging schedule is demonstrated to improve distributed network management. In reference [80], a storage unit is owned by a wind farm and its operational decisions are established according to the wind farm's power production estimates in order to maximize the owner's profit. In reference [81], a storage unit is operated by an independent utility to maximize its profit obtained in both energy and ancillary markets.

However as argued in reference [14] and reference [65], naïve application of dynamic pricing fails to capture the actual value of FD and ES flexibility, as they discard the natural interdependence between FD/ES and market prices, as the communicated price signals are not affected by the resulting demand response close to real time[25], [27]. Without such feedback of FD's and ES's operation on prices, inefficient or even infeasible market outcomes[14] could emerge. For instance, if a large number of flexible consumers are exposed to dynamic prices, clearly all the consumers are motivated to shift the operation of their loads towards the lowest-priced periods, creating new (possibly considerable) demand peaks, and the economic implications of which are not accounted for in the published prices. To deal with such demand response concentration challenge, the authors in reference [65] impose heuristic flexibility restriction on flexible loads, in order to restrain consumers from concentrating their demand at the lowest-priced periods. The same authors in references [82]—[83] further suggest means of penalizing the extent of flexibility utilized by flexible loads through a flexibility price and/or randomizing the prices transmitted to different flexible loads, so as to diversify their response and limit the concentrated shift of re-schedulable demand to lowest-priced periods.

2.4 Participation of Flexible Demand and Energy Storage in Different Market Segments

As discussed in Section 1.2, FD and ES have the potential to provide multiple services to several sectors in electricity industry and thus support activities related to generation, network, and system operation. At the same time, these split benefits of FD and ES present significant challenges to design adequate market mechanisms that ensure investors in these technologies to be properly rewarded for delivering these various sources of value. In this context, apart from participating in the energy market, FD and ES are envisaged to simultaneously participate in markets for the provision of system services (such as balancing and network services) in order to realize their full flexibility potential. Previous work pertaining to the multi-service provision of FD and ES in electricity markets is briefly discussed in this section. Reference [13] and reference [19] quantify the system wide benefits of FD and ES across multiple sectors of electricity (including generation, transmission, and distribution), based on a whole-system cost minimization approach. A co-optimized compressed air energy storage (CAES) dispatch model is developed in reference [84] to quantify the value of providing operating reserves in addition to energy arbitrage. It is demonstrated that energy arbitrage and operating reserves net revenues could make CAES units profitable in electricity markets, therefore support their investments. A security-constrained market clearing mechanism considering FD participation and provision of system reserves is proposed in reference [85]. The results demonstrate that the system operator could exploit demand flexibility to shift generation reserve scheduling and deployment from high- to low- priced periods, leading to reduced operational costs. It is also suggested that if the demand side exhibits high flexibility, it could compete with generators in providing system reserves.

While references [13], [19], [84]—[85] have analysed the diverse sources of value that the participation of FD and ES promise to the electricity system, references [86]—[88] focus on devising suitable strategies for optimizing FD/ES multi-service portfolios in different market segments, under different market designs. In reference [86], an optimization problem is formulated to schedule operation of distributed ES by coordinating provision of a range of system services, which are remunerated at different market prices. The model maximizes the profit of ES while providing energy arbitrage, and various reserve and frequency regulation services. The model facilitates the participation of ES in the markets for the provision of balancing services through selecting of the most profitable multi-service portfolio. The results demonstrate that under high energy price volatility, ES derives higher profit in the energy market and thus commits less in the balancing markets and vice versa. The same authors in reference [87] identify that the multi-service provision of ES could lead to battery aging and degradation, the resultant impact on ES short-term and long-term profitability is examined. A profit maximization problem of an EV aggregator providing both energy and frequency regulation services through Vehicle-to-grid (V2G) control is examined in reference [88]. A dynamic programming algorithm is developed for the aggregator to determine the optimal charging strategy of individual EV. Simulation results are provided to illustrate the optimality of the proposed V2G control strategy with sensitivity analysis performed regarding different EV parameters. References [86]—[88] provide insights associated with the development of market arrangements that are capable of adequately incentivize FD and ES for their multi-service provision.

2.5 Centralized Market Clearing Mechanism with Flexible Demand and Energy Storage Participation

As discussed in Section 2.3, the centralized mechanism achieves active participation of FD and ES in electricity markets. Under such mechanism, the optimal operational decisions of FD and ES are established on a global efficiency target to maximize social welfare. Therefore, it enables to clearly determine the potential benefits obtained by integrating FD and ES into the system operation. Moreover, the centralized mechanism takes into account the effect that FD and ES may have in the market operation, particularly, their contributions to the market price-setting process. In other words, it also enables to analyse the impact of FD and ES on the market clearing outcome, and subsequently the impact on other market participants' behaviour. Given the focus of this monograph is on analysing the integration of FD and ES in electricity markets, along with the aforementioned desirable properties, the centralized market clearing mechanism is employed throughout the main chapters (Chapter 3 — Chapter 5) of this monograph. This section firstly outlines the main assumptions made regarding the design and participants' behaviour of the examined electricity markets, and then presents a basic mathematical formulation of the centralized market clearing mechanism.

2.5.1 Assumptions on Examined Electricity Market Model

The main assumptions made in this monograph concerning the examined electricity markets are discussed in the following subsections.

1. *Traded commodity*. Market arrangements in deregulated power systems may involve the market-oriented provision of different commodities, such as electrical energy, capacity, and a variety of ancillary services[46]. In this monograph, electrical energy is assumed to be the only commodity that is traded in the examined market. As discussed in Chapter 6 however, a future work direction will investigate the potential of capacity and ancillary services provided by flexible demand and energy storage, and analyse the impact of their participation at different market segments, under alternative market designs.

2. *Type of the markets*. Among the market models implemented in different parts of the world, it is possible to distinguish two main types of market organization: bilateral contracts and pool-based markets[46]. In the former, producers and consumers

independently enter into bilateral contracts, where the producers agree to provide energy service, while the consumers agree to a price schedule for this service. Pool-based markets are assumed in this monograph, where the participants do not directly interact with one another, but only indirectly through a central entity referred to as the market operator. Generation/demand participants submit their offers/bids to the market operator, respectively, specifying the price and quantity at which they are willing to sell/buy energy. On the basics of the collected offers and bids, the market operator incorporates a systematic settlement mechanism to determine the market equilibrium comprising of the clearing quantities (i.e. the dispatch of each participant) and the clearing prices.

Following the common practice of the pool-based market models presented in the literature and implemented around the world, the objective of the market clearing mechanism is set as the maximization of the social (global) welfare[72]. In the context of electricity markets, the social welfare quantifies the overall benefit that arises from energy trading. Specifically, it is defined as the difference between the benefit consumers perceive from consuming the energy and the cost producers incur from producing them[46]. As such, it also represents the sum of the participants' surplus at the prevailing market clearing prices. The producers' surplus or producers' profit is defined as the difference between the revenue it receives from the sale of the energy it produces, and the cost of producing this energy, while the consumers' surplus or consumers' utility is defined as the difference between the consumer benefit, and the payment for the energy purchased[46], [72].

3. *Market horizon and temporal resolution.* Depending on the market horizon, a pool-based market can be organized as day-ahead (the trading of energy takes place on one day for the delivery of electricity the next day), hourly-ahead, etc. A relatively long market horizon may bring participants greater uncertainty in predicting their generation/demand parameters very long into the future. On the other hand, the FD and ES participants examined in this monograph are characterized by an ability to redistribute energy demand in time (subsection 1.2.1). In this context, a relatively short horizon may not capture sufficiently participants' inter-temporal behaviour[14]. Following the common practice of most market designs, a day-ahead horizon is employed in this monograph, which constitutes a good compromise between the long and short horizons.

Regarding the selection of the temporal resolutions within the market horizon, a higher resolution improves representation of participants' characteristics, however, presents computational challenges. The examined market model therefore involves an hourly resolution, which has been historically acceptable for electricity market modelling. It captures the basic properties of most generation and demand participants[14] and at the same time demands moderate computational power.

4. *Pricing mechanisms*. The pricing mechanism plays a vital role in the clearing mechanism adopted by the market operator. The widely adopted mechanism in most electricity market designs is *marginal pricing*, where the price is set as the marginal generation cost (or marginal social welfare in general) of serving an additional unit of demand. Mathematically, the Lagrangian multipliers associated with system demand-supply balance constraint represent the marginal prices. These prices are uniform for all market participants and determine the payment made from/to the consumers/producers. However, as thoroughly discussed in Chapter 3, uniform marginal pricing cannot generally support competitive equilibrium solutions in markets with non-convexities. To address this challenge, the market operator generally adopts more advanced pricing mechanisms. Among others, these include alternative uniform pricing, non-linear, differentiated pricing in conjunction of various types of financial compensation schemes[34-45].

5. *Network effect and Nodal pricing*. In realistic power systems, market participants are located in distinct geographical areas attached to different nodes and connected by transmission lines in the electric power network. The network constraints therefore must always be respected for the sake of the secure operation of the system. In this monograph, the model used to account for the effect of the network over which the market operates is based on the direct current optimal power flow (DCOPF) model[46].

Extending the aforementioned concept of marginal pricing to account for the network effect, the centralized market clearing mechanism examined in this monograph employs the *locational marginal pricing* scheme, where the price at a specific location (node) of the network is determined as the marginal generation cost of serving an additional unit of demand at that location, without violation of the network constraints[46]. If the network is entirely uncongested, the prices are identical at all nodes, reflecting the marginal cost of serving the next increment of system demand. In cases of transmission congestion, on the other hand, the prices differ by location, since the congestion prevents energy from low-cost generation from meeting demands at all locations. As such, some of the low-cost generation needs to be curtailed, lowering the market price in sending area of the congestion; the output of the higher-cost generators must be increased to serve demand (and thus raising the market prices) in the receiving area of the congestion.

In Chapter 3, the network constraints are neglected in the market model due to the desirable computational simplicity. In addition, the fundamental effects: schedule inconsistency and surplus sub-optimality in market with non-convexities can be demonstrated clearly even in a single-node system. Moreover, as discussed in subsection 3.5.3, an equitable distribution rule of the total surplus loss compensation among market participants is proposed in the extended generalized uplift approach. However, the concept of "fairness" becomes ambiguous to define and justify when the network effects are accounted for, as it may be unfair for participants located at different nodes of the network to contribute to compensate each other's surplus loss. As discussed in subsection 6.2.1, however, further work will focus on extending the pricing mechanisms (developed in Chapter 3) to include the network representation so as to produce appropriate locational incentives in form of prices/uplift payments, and also extending the proposed equitable surplus loss sharing rule to achieve fairness in network constrained electricity markets.

6. *Market competition*. The modelling methodologies required to represent and examine the operation of electricity markets differ according to the assumed behaviour of the participants. In this regard, market structure can be classified into two categories: markets with *perfect competition* and markets with *imperfect competition*[46].

In a perfectly competitive market, participants are assumed to submit offers/bids to the market revealing their truthful techno-economic characteristics. Participants act as price-takers, that is, no participant has the ability to influence the market price through its individual actions. Such assumption is valid if the number of participants is vast and the volume handled by each participant is small compared to the overall market volume. However, this assumption does not hold in the case of the deregulated electricity markets[46].

In a market with imperfect competition, on the other hand, some participants owning a large share of the market are able to manipulate market prices and increase their surplus beyond the competitive equilibrium levels, through strategic offering/bidding (i.e. not revealing their actual characteristics to the market). This exercise of market power, however, is achieved at the expense of the social welfare. Depending on the number of strategic participants, markets with imperfect competition can be further divided into two extremes: monopoly and oligopoly. In the former, there is a sole monopolist who can set the market price at will. In the latter, at least two participants (controlling a large market share) act strategically to exercise market power, other participants with a small market share act as *competitive fringe* (price-takers). Due to the specific features of the electricity industry (such as only a limited number of competing generation firms), the deregulated electricity markets are better described as oligopoly, and such models have become a natural framework to study markets with imperfect competition[89].

It is worth stressing that the optimization models representing perfect and imperfect markets are fundamentally different. In order to represent perfect markets, it suffices to adopt a single optimization problem representing the market clearing problem (solved by the market operator) that determines the dispatch of each participant to maximise the social welfare (given the truthful bids of the participants). Modelling and analysing the operation of imperfect markets, on the other hand, presents more challenges. The market clearing problem is formulated based on the perceived misreported participants' economic and technical characteristics, leading to different objective functions or constraints of the problem. The impact of the strategic actions of participants on the market clearing outcomes therefore needs to be incorporated in the model. In addition, due to the fact that more than one participant can alter the market outcome; each participant has to consider the possible actions of its rivals when selecting its strategy.

This monograph models both electricity markets with perfect and imperfect competition. However, given that the focus of Chapter 3 is on developing pricing mechanisms to address FD non-convexities, it is assumed that the examined market is perfectly competitive. As mentioned in Chapter 6 however, a direction of further work will involve examining the incentive compatibility of the developed pricing mechanisms. Furthermore, the participation of FD and ES in markets with imperfect competition is comprehensively explored in Chapters 4 and 5. Chapter 4 examines the impact of FD and ES on the market power exercised by the generation side, while Chapter 5 focuses on exploring the potential of market power exercised by the FD and ES participants.

2.5.2 Model of Centralized Market Clearing Mechanism

As discussed in subsection 1.2.2, pool energy markets have traditionally employed the centralized clearing mechanisms for FD and ES integration. This subsection presents a general model of such a mechanism under the assumptions made in the previous section. Market participants submit their offers/bids (reflecting information regarding their individual economic and technical characteristics) to the pool market operator. Using this collected knowledge, the latter derives the participants' cost/benefit functions and sets of operational constraints and determines the clearing dispatch and prices through a social welfare exercise over the considered day-ahead horizon. This market clearing optimization problem is formulated as:

$$\min_{V} \sum_i C_i(\boldsymbol{g}_i) - \sum_j B_j(\boldsymbol{d}_j) \tag{2.1}$$

where:

$$\boldsymbol{V} = [\boldsymbol{g}_i, \forall i] \cup [\boldsymbol{d}_j, \forall j] \cup [\boldsymbol{s}^c, \boldsymbol{s}^d] \cup [\boldsymbol{\theta}_t] \tag{2.2}$$

subject to:

$$e_{n,t}(\boldsymbol{g}_i, \boldsymbol{d}_j, \boldsymbol{s}^c, \boldsymbol{s}^d, \boldsymbol{\theta}_t) = 0, \forall n, \forall t \tag{2.3}$$

$$F_{n,m,t}(\theta_{n,t}, \theta_{m,t}) \leqslant F_{n,m}^{\max}, \forall n, \forall m \in M_n, \forall t \tag{2.4}$$

$$\boldsymbol{g}_i \in G_i, \forall i \tag{2.5}$$

$$\boldsymbol{d}_j \in D_j, \forall j \tag{2.6}$$

$$\boldsymbol{s}^c, \boldsymbol{s}^d \in S \tag{2.7}$$

where:

t: Index of time periods.

i: Index of generation participants.

j: Index of demand participants.

n, m: Index of nodes.

M_n: Set of nodes connected to node n.

$\boldsymbol{g}_i$: Vector of hourly electric power productions $g_{i,t}$ of generation participant i.

$\boldsymbol{d}_j$: Vector of hourly electric power demand outputs $d_{j,t}$ of demand participant j.

$\boldsymbol{s}^c$: Vector of hourly charging power s_t^c of energy storage participant.

$\boldsymbol{s}^d$: Vector of hourly discharging power s_t^d of energy storage participant.

$C_i(\cdot)$: Daily cost function of generation participant i.

$B_j(\cdot)$: Daily benefit function of demand participant j.

G_i: Operational constraint set of generation participant i.

D_j: Operational constraint set of demand participant j.

S: Operational constraint set of energy storage participant.

$\boldsymbol{\theta}_t$: Vector of hourly voltage angle $\theta_{n,t}$ at node n.

$e_{n,t}$: Hourly system demand-supply power imbalance at node n.

$F_{n,m,t}$: Hourly power flow on transmission link connecting node n and node m.

$F_{n,m}^{\max}$: Capacity of transmission link connecting node n and node m (MW).

This problem is subject to two groups of constraints: system-level and local-level constraints. The former involve constraints coupling all market participants and include the coordination of the hourly demand-supply balance (2.3) and hourly network transmission link's capacity constraint (2.4). The latter involve constraints corresponding to the technical and economic operational characteristics of individual participants (2.5)—(2.7). The values of the decision variables at the optimal solution constitute the market clearing schedules for the participants, and the Lagrangian multipliers of (2.3) constitute the locational marginal prices (LMP).

It should be noted that the above basic framework for the centralized market clearing mechanism is adopted throughout the main chapters of this monograph. However, the assumption of the examined market and detailed modelling of different market participants may differ in each chapter. These are briefly discussed next.

Chapter 3: Given that the focus of this chapter is to identify non-convexities associated with the market participants, the local-level constraints consider the latter's inter-temporal operational characteristics, and the mathematical modelling of which generally requires the use of integer/binary variables. For example, the minimum stable generation constraint of a generation participant is related to the commitment statues (binary variables) of the participant (subsection 3.3.1); similarly for the minimum and maximum power limits of a CCFD participant (subsection 3.3.2). Energy storage participants are not examined in this chapter. Furthermore, after solving the centralized market clearing problem (which corresponds to a mixed-integer program due to the existence of the integer/binary variables), a continuous version of which is solved subsequently with the binary variables set equal to their optimal values. The Lagrangian multipliers associated with the demand-supply balance constitute the electricity prices. This procedure is adopted as Lagrangian multipliers cannot be directly calculated from the solution of problems in the presence of binary variables (subsection 3.3.3).

Chapter 4 and Chapter 5: These two chapters focus on markets with imperfect competition, and the modelling of which requires a bi-level optimization problem that captures the interaction between strategic participants and the market operator (subsection 2.5.1). In Chapter 4, individual strategic electricity producer constitutes the player in the upper level (UL) problem of the bi-level model. In Chapter 5, the UL problem represents the decision-making of strategic ES participant. The lower level (LL) problem in both chapters represents the market clearing process which endogenously models participants' actions on the market clearing outcome. In order to solve this bi-level problem, the LL problem is replaced by its optimality conditions (subsection 4.5.2), which is enabled by the *continuity* and *convexity* of the LL problem. It is therefore assumed that all market participants are without non-convex operational characteristics. Furthermore in both chapters, the own-price elasticity and the time-shifting flexibility of the demand side are captured in the proposed operational models, expressing the consumers' ability to redistribute their energy requirements across time without compromising the satisfaction they experience.

Chapter 3
Factoring Flexible Demand Non-Convexities in Electricity Markets

3.1 Introduction

The deregulation of the electricity markets has witnessed worldwide over the last three decades, replacing the centralized operation paradigm with a market-oriented framework. In this context, a concern arises as to whether the market arrangements will provide appropriate incentives to the self-interested market participants to deliver the social welfare maximizing system operation solutions.

As the economic literature has demonstrated, in electricity markets with non-convexities, uniform linear pricing cannot generally support competitive equilibrium solutions[33-45]. In other words, individual market participants' surplus maximizing self-schedule given the prices determined by the centralized market clearing problem, is not generally consistent with the schedule calculated by the latter. In cases of such inconsistencies, the centralized schedules entail lower surpluses than self-scheduling, with this difference termed as surplus loss and the related effect as surplus sub-optimality. This effect is undesirable on the basis of allowing all self-interested market participants to determine independently their position given the prices, and not by central intervention, and adversely influencing the fairness and efficiency of the markets. Furthermore, in the long run, forcing participants operating with surplus sub-optimality will drive them out of business, and will also discourage necessary investments, endangering the security of the system.

A wide range of literature has identified non-convexities (see Appendix A for a brief discussion on convexity principles) associated with the generation side of electricity markets, including binary (on/off) commitment decisions, fixed and start-up/shut-down costs, minimum stable generation constraints, and minimum up/down times[34-45]. As demonstrated in the literature, these non-convexities lead to the generators' schedules inconsistency and profit sub-optimality effects. Over time, different mechanisms[34-45] for pricing in electricity markets with generation non-convexities have been explored to address such effects. Among others, these include uniform pricing and non-linear, differentiated pricing and incorporation with different types of financial compensation schemes (Section 3.2).

Regarding the demand side of the market, in line with the emerging *Smart Grid* paradigm[56], recent developments have paved the way for the introduction of flexible demand (FD) in power systems, with significant economic, technical, and environmental benefits[13-18]. In the competitive environment, the realization of this potential is coupled with the integration of FD in electricity markets[21]. Authors in references [14], [22]—[27], [65]—[67] have proposed different market clearing mechanisms considering FD participation and demonstrated the impact of FD on the market. However, previous work has not explored non-convexities and surplus sub-optimality effects associated with FD.

This chapter develops detailed operational models for different types of FD capturing their distinct flexibility potential, and identifies for the first time non-convexities associated with the operation of FD and demonstrates their relation with inconsistency and surplus sub-optimality effects through simple examples and a large case study with day-ahead horizon and hourly resolution. Previously proposed pricing mechanisms addressing generation non-convexities are adequately extended to account for FD participation in electricity markets.

This chapter is organized as follows. Section 3.2 conducts a comprehensive literature review on previously proposed pricing mechanisms in electricity markets with generation non-convexities. Section 3.3 derives detailed operational models of FD and formulates the centralized market clearing problem under FD participation. Section 3.4 identifies FD non-convexities and demonstrates the resulting schedules inconsistency and surplus sub-optimality effects through simple examples. Section 3.5 and 3.6 detail the extension of generalized uplift and convex hull pricing approaches respectively to account for both generation and FD participation. Section 3.7 presents the examined case study and Section 3.8 discusses conclusions of this chapter.

3.2 Literature Review

As discussed in the previous section, to unburden the undesirable schedule inconsistency and profit sub-optimality effects that non-convexities may create in the market, it necessitates the market operator to employ alternative pricing mechanisms that adequately incentivize market participants. Two general approaches have been explored to address such effects.

Characterized by simplicity and transparency, the first approach retains uniform or linear pricing and attempts to minimize the extent of schedule inconsistency and profit sub-optimality. A *primal-dual* formulation of the centralized dispatch problem is proposed in reference [34] in order to determine the electricity prices minimizing the social welfare reduction caused by the schedules inconsistency, while ensuring profit non-negativity for the generators. Authors in references [35]—[36] proposed an alternative pricing approach for the day-ahead electricity market, the approach employs a *Semi Lagrangian Relaxation* (SLR) technique and a second price auction. The second price auction provides the final electricity price for the day-ahead electricity that ensures that the engaged generators are break even. Although schemes in references [34]—[36] guarantee the revenue adequacy of the generators, they do not achieve competitive equilibrium at the optimal solution of the centralized problem, and does not guarantee zero profit loss for the generators as it does not ensure recovery of opportunity costs.

It has been widely noted that in the presence of non-convexities there may be no uniform prices that support market equilibrium. To this end, in references [37]—[40], apart from the market-based payment (derived from uniform prices), generators experiencing profit sub-optimality also receive uplift payments that compensate exactly their respective profit loss, and the uniform prices are optimized so as to minimize the total profit loss and thus the total uplift payments. These minimum-uplift prices correspond to the *convex hull prices* and coincide with the Lagrangian multipliers optimizing the dual of the market clearing problem[38-40]. The compound of convex hull prices and minimum uplift payments ensure the generators indifferent between accepting the centralized solution and responding optimally to the prices.

The second approach addresses schedules inconsistency and profit sub-optimality by employing additional, *differentiated* (i.e. generator-specific) prices. In reference [41], after solving the initial mixed-integer centralized problem, authors solve a continuous version of the latter, with the binary commitment variables set equal to their optimal values. The dual variables of these equality constraints yield differentiated prices for the generators' commitment (*commitment tickets*), which along with the uniform energy prices to support an equilibrium solution. However, these prices may display undesirable properties of high volatility. Recognizing this, authors in reference [42] propose an alternative non-uniform pricing rule where equilibrium prices compounded of a more stable commodity price and an uplift charge are determined based on the generation of a separating valid inequality that supports the optimal energy allocation. However, approaches in references [34]—[42] entail that the total compensation of profit loss is entirely charged to the demand side of the market, which is thus treated inequitably. To this end, authors in references [43]—[45] propose the use of *generalized uplift functions*, which include generator-specific linear and non-linear terms and constitute additional revenues or payments for the generators. The parameters of these functions along with the electricity prices are adjusted to achieve consistency for each generator and an equitable distribution of the generators' profit loss compensation among the market participants.

Although the generalized uplift approach yields more equitable distribution of the total surplus loss compensation, it introduces price discrimination among the market participants that cannot be easily justified and may be considered non-transparent[33-34], [37]. For this reason, the calculation of these differentiated prices in reference [44] is carried out through an optimization problem minimizing the extent of discrimination introduced. Since the approach developed in reference [44] could not efficiently deal with multi-period problems accounting for generators' time-coupling characteristics, an iterative cutting-plane algorithm for the calculation of uplift parameters and electricity prices is proposed in reference [45].

In this chapter, both the generalized uplift and convex hull pricing approaches addressing generation side's schedule inconsistency and surplus sub-optimality effects are suitably extended to account for FD non-convexities. Concerning the former, generalized uplift functions for FD participants are proposed, which include FD-specific terms and constitute additional benefits or payments for the FDs. The structure of generalized uplift functions for FDs is not the same with the respective function for generators, due to their different non-convexities and resulting surplus sub-optimality effects. The parameters of these functions along with the electricity prices are adjusted to achieve consistency for each FD participant. A new rule is introduced for equitable distribution of the total generators' profit loss and FDs' utility loss compensation among market participants. Regarding the latter, convex hull prices are calculated as the Lagrangian multipliers optimizing the dual problem of the market clearing problem considering FD participation. It is demonstrated that convex hull prices are flattened at periods when FD is scheduled to eliminate surplus sub-optimality associated with the FD ability to redistribute energy requirements across time.

3.3 Centralized Market Clearing Under Flexible Demand Participation

3.3.1 Modelling Generation Participants

The cost function that corresponds to a generation participant i is expressed as equation (3.1) consisting of a *quadratic variable cost* component that is associated with the level of power it generates; a *fixed cost* component that is associated with the unit commitment status, and is incurred when the unit is on; a *start-up* and a *shut-down cost* component that are incurred when the unit is put online or brought offline (they are modelled through constraints (3.5)—(3.8)).

$$C_i(\xi_i) = \sum_t a_i u_{i,t} + b_i g_{i,t} + c_i g_{i,t}^2 + SU_{i,t} + SD_{i,t} \tag{3.1}$$

The operating constraints set of a generation participant i includes the following constraints:

When the generator is online, its power output is bound by a minimum and a maximum power limit.

$$u_{i,t} g_i^{\min} \leqslant g_{i,t} \leqslant u_{i,t} g_i^{\max}, \ \forall t \tag{3.2}$$

The generator is constrained on upward and downward output variation by rates R_i^u and R_i^d, respectively:

$$g_{i,t} - g_{i,(t-1)} \leqslant R_i^u, \ \forall t \tag{3.3}$$

$$g_{i,(t-1)} - g_{i,t} \leqslant R_i^d, \ \forall t \tag{3.4}$$

A constant start-up cost C_i^u or a shut-down cost C_i^d is incurred only when the unit is brought online or put offline, respectively:

$$SU_{i,t} \geqslant C_i^u [u_{i,t} - u_{i,(t-1)}], \ \forall t \tag{3.5}$$

$$SU_{i,t} \geqslant 0, \ \forall t \tag{3.6}$$

$$SD_{i,t} \geqslant C_i^d [u_{i,(t-1)} - u_{i,t}], \ \forall t \tag{3.7}$$

$$SD_{i,t} \geqslant 0, \ \forall t \tag{3.8}$$

It is required that when the unit is scheduled on or off, it has to remain in that status for a certain amount of time before it can change status again. The minimum required up and down times are denoted by UT_i and DT_i, respectively. It follows the method proposed in reference [90], the minimum up and down time constraints are formulated as mixed-integer linear expressions based solely on the unit commitment decisions $u_{i,t}$. The minimum up/down time constraints are expressed as:

$$\sum_{t=1}^{H_i^u}[1-u_{i,t}]=0 \qquad (3.9)$$

$$UT_i[u_{i,t}-u_{i,(t-1)}] \leqslant \sum_{n=t}^{t+UT_i-1} u_{i,n}, \quad \forall t=H_i^u+1,\cdots,N_T-UT_i+1 \qquad (3.10)$$

$$\sum_{n=t}^{N_T}[u_{i,n}-(u_{i,t}-u_{i,(t-1)})] \geqslant 0, \quad \forall t=N_T-UT_i+2,\cdots,N_T \qquad (3.11)$$

$$\sum_{t=1}^{H_i^d} u_{i,t}=0 \qquad (3.12)$$

$$DT_i[u_{i,(t-1)}-u_{i,t}] \leqslant \sum_{n=t}^{t+DT_i-1}[1-u_{i,n}], \quad \forall t=H_i^d+1,\cdots,N_T-DT_i+1 \qquad (3.13)$$

$$\sum_{n=t}^{N_T}[1-u_{i,n}-(u_{i,(t-1)}-u_{i,t})] \geqslant 0, \quad \forall t=N_T-UT_i+2,\cdots,N_T \qquad (3.14)$$

where parameters H_i^u and H_i^d represent the number of initial periods that unit i must be online and offline respectively, which can be mathematically expressed as (3.15)—(3.16).

$$H_i^u = \min\{N_T, [UT_i-U_i^0]u_{i,0}\} \qquad (3.15)$$

$$H_i^d = \min\{N_T, [DT_i-D_i^0][1-u_{i,0}]\} \qquad (3.16)$$

Constraints (3.9) and (3.12) are related to the initial status of the units as defined by H_i^u and H_i^d, respectively. Constraints (3.10) and (3.13) enforce that in subsequent periods (of size UT_i and DT_i) minimum up and down time constraints are, respectively, satisfied. Constraint (3.11) models the final UT_i-1 periods in which if unit i is brought online, it remains online until the end of the time horizon. Analogously, constraint (3.14) enforces that already-offline unit i remains offline until the end of time span if required by its minimum down time constraints.

3.3.2 Modelling Flexible Demand Participants

Flexible demand (FD) generally refers to any type of consumer who has the flexibility to alter the electricity usage from its typical consumption pattern in response to changes in price of electricity over time through the employment of some forms of storage[16]. FD participants may generally correspond to large industrial/commercial consumers, participating individually in the market, or FD aggregators, representing a large number of smaller flexible residential/commercial consumers (references [75]—[76]). FD generally includes assorted domestic and commercial loads as well as quantity of industrial loads that can be operated flexibly[14]. However, the derivation of a strictly accurate model of the operation of the FD technologies according to the whole range of their technical specifications is out of the scope of this monograph.

Nevertheless, two different types of FDs are considered in this chapter, capturing the largest part of flexible load models in the related literature: *continuously-controllable* FD (CCFD) and *fixed-cycle* FD (FCFD)[91]. The power demand of an FD of the first type can be continuously adjusted between a minimum and a maximum limit when the FD is active (i.e. its demand is not zero). The FD of the second type involves operating cycles which comprise a sequence of phases occurring at a fixed order, with fixed duration and fixed power consumption, which cannot be altered. Without loss of generality, it is assumed that a demand activity of an FCFD corresponds to one such fixed cycle, and that this cycle cannot be interrupted once it is initiated.

Furthermore, two different types of demand flexibility potential are considered. The first is associated with the ability to completely forgo or abandon demand activities[15], [92]. In order to account for the flexibility potential, the benefit function associated with FD j is expressed by equation (3.17).

$$B_j(\psi_j) = B_j^0 \times v_j \qquad (3.17)$$

"Forgoing demand activities" implies that the operation of an electrical load planned by the consumers is not carried out[15], [92]. For example, in the context of a domestic demand, the operation of a planned washing machine operation is not carried out. Moreover, the execution of the electrical load operation generally provides some benefit or "usefulness" to the consumers. For instance, the benefit of a washing machine operation is the satisfaction perceived by the consumer due to the availability of clean clothes. In this context, the benefit FD participants perceive is modelled as a function of the binary decision expressing whether the related demand activity is forgone or not. More specifically, when the demand activity of FD j is forgone, then the respective binary variable v_j is equal to zero, and the benefit of the respective consumer is equal to zero. When the demand activity of FD j is not forgone (i.e. the related load operation is executed), then the respective binary variable v_j is equal to one, and the benefit of the respective consumer is equal to a positive value B_j^0, which depends on the extent of usefulness perceived by the consumers when the demand activity is carried out.

The second type of flexibility potential is associated with the ability to redistribute the total electrical energy requirements of activities across time[15]. Without loss of generality, it is assumed that each FD participant carries out at most one activity over the market horizon.

Regarding the CCFDs, they are characterized by the flexibility regarding the specific time period that they can acquire the amount of energy needed for the operation of their loads, as long as the acquisition is carried out within a user-specified temporal interval[91]. In addition, the flexibility is also associated with continuously controllability of their power demand at each time period within this interval, as long as it is within the minimum and the maximum power limits. *Electric vehicle* (EV) with flexible charging capability constitutes a representative example of a CCFD.

Figure 3.1 illustrates the flexibility characteristics associated with the operation of a CCFD. The solid and the dashed power profiles correspond to two different demand patterns. Both consumption patterns make sure that the total energy required for the operation of the load is acquired within the scheduling period, but the timing and power level of this energy input is different. It thus can be concluded that the CCFDs' extent of operational flexibility depends on the length of their scheduling period and the relative size of their maximum power limit with respect to their overall energy requirement.

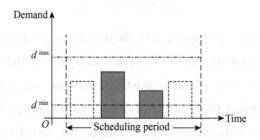

Figure 3.1 Example of flexibility exploitation of a CCFD

Mathematically, the operating constraints set D_j of a CCFD j includes the following constraints:

- The total energy consumption is zero if the demand activity is forgone, or equal to the fixed energy requirement of the activity otherwise:

$$\sum_t d_{j,t} \times 1h = v_j \times E_j \qquad (3.18)$$

- When the CCFD is active, its power demand is bounded by a minimum and a maximum power limit:

$$w_{j,t} d_{j,t}^{\min} \leqslant d_{j,t} \leqslant w_{j,t} d_{j,t}^{\max}, \quad \forall t \qquad (3.19)$$

As revealed in constraints (3.18)—(3.19), the power demand of a CCFD j can be continuously adjusted within the range of the above minimum and maximum power limits. However, they cannot take any values between 0 and minimum power limit, namely $d_{j,t} \in \{0\} \cup [d_{j,t}^{\min}, d_{j,t}^{\max}]$.

FCFDs exhibit flexibility regarding the timing that they can acquire the required amount of energy within the scheduling period allowed by their users[91]. However, by contrast the CCFDs, they are not flexible in terms of their power demand at each time period. This is due to the fact that FCFDs do not incorporate explicit storage components (such as the battery of EV) and therefore the acquisition of the required electrical energy and its actual consumption remain temporal coupled. Additionally, their operations involve operating cycles which comprise a sequence of phases occurring at a fixed order, with fixed duration and fixed power consumption, which cannot be altered. As a result, their flexibility is only associated with their ability to postpone these cycles in time within the scheduling period. Wet appliances (including washing machines, dish-washers, tumble dryers, etc.) with deferrable initiation time represent well the operation of the FCFDs.

The flexibility characteristics associated with the operation of an FCFD are illustrated in Figure 3.2, where the solid and the dashed power profile corresponds to two different consumption patterns. Both of them ensure that the appliance cycle is executed following the inherent power profile, but the activation timing of this cycle executed distinctly. Therefore, the FCFDs' extent of operational flexibility depends only on the length of their scheduling period.

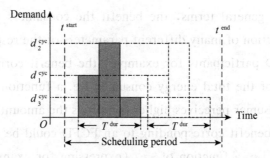

Figure 3.2 Example of flexibility exploitation of an FCFD

Mathematically, the operating constraints set D_j of an FCFD j includes the following constraints:

- The demand activity can be executed over the time window determined by t_j^{start} and t_j^{end}:

$$z_{j,t}=0; \quad \forall t < t_j^{\text{start}} \text{ and } \forall t > t_j^{\text{end}} - T_j^{\text{dur}} + 1 \qquad (3.20)$$

- The demand activity is executed (and thus initiated) at most once during the above time window:

$$\sum_{t=t_j^{\text{start}}}^{t_j^{\text{end}} - T_j^{\text{dur}}+1} z_{j,t} = v_j \qquad (3.21)$$

- If the activity is not forgone, the demand at each period depends on the initiation time, T_j^{dur} and $d_{j,\tau}^{\text{cyc}}$; $\forall \tau \in [1, T_j^{\text{dur}}]$:

$$d_{j,t} = \sum_{\tau=1}^{T_j^{\text{dur}}} z_{j,(t+1-\tau)} \, d_{j,\tau}^{\text{cyc}}, \quad \forall t \qquad (3.22)$$

Similarly to CCFDs, operational constraints (3.20)—(3.22) suggest that once an FCFD j is activated, its demand $d_{j,t}$ can only take a value out of a set of discrete values (as defined by the power requirement of each step of the operation cycle), i.e. $d_{j,t} \in \{0, d_{j,1}^{\text{cyc}}, d_{j,2}^{\text{cyc}}, \ldots, d_{j,T_j^{\text{dur}}}^{\text{cyc}}\}$.

For the sake of generality, both CCFDs and FCFDs are assumed to exhibit both flexibility potential (i.e. forgo their demand activities and redistribute the energy requirements of activities across time). However, straightforward modifications can be carried out to model FDs with the potential. If for example the activity related to an FD cannot be forgone, its respective binary variable v_j will be forced equal to 1. Moreover, an FCFD that cannot redistribute its activity will be modelled with $t_j^{\text{start}} = t_j^{\text{end}} - T_j^{\text{dur}} + 1$. It is worth noting that the proposed FD models can easily be expanded to form a more accurate representation of flexible loads operation.

Benefit function: In general terms, the benefit the consumers perceive by a demand activity can be a function of many different parameters of the respective load operation. In the context of FD participants for example, the benefit corresponding to a CCFD could be a function of the total energy consumed i.e. a function of E_j (expressing for example that the consumer perceives higher benefit as the amount of consumed energy is increased), and the benefit corresponding to an FCFD could be a function of the time the cycle is initiated i.e. a function of $z_{j,t}$ (expressing for example that the consumer perceives higher benefit if the cycle is executed earlier in the day). Although such more complex benefit functions may capture more accurately the preferences of some consumers, such functions do not introduce additional FD non-convexities (subsection 3.4.3) and thus are not included in the model for simplicity and brevity.

Operational constraints: Additional constraints can be included in the proposed FD models to capture some practical aspects of FD operation. For example, constraints on the delay between consecutive activation, constraints on the frequency of activation, etc. Take a CCFD j for instance, constraint (3.23) below expresses a limit on the maximum frequency of activation; the total number of periods that the CCFD is active cannot be higher than a maximum limit n_j:

$$\sum_t w_{j,t} \leqslant v_j \times n_j \qquad (3.23)$$

while constraint (3.24) below expresses that the delay between two consecutive activations of the CCFD cannot be lower than a minimum limit $del_j^{\min}$.

$$\sum_{\tau=1}^{del_j^{\min}+1} w_{j,(t+\tau-1)} \leqslant 1, \ \forall t \qquad (3.24)$$

Although the above delay and frequency constraints represent practical aspects of some flexible load operation, they do not contribute to additional FD non-convexities (subsection 3.4.3) and thus are not included in the model.

3.3.3 Centralized Market Clearing Solutions

As discussed in subsection 2.5.2, pool energy markets employ a centralized market clearing mechanism, where the market operator derives the participants' cost/benefit functions and operational constraints on the basics of their respective bids/offers, and determines the centralized solutions (constitute of market clearing dispatch of the participants and clearing prices) by solving a social welfare maximization problem over the day-ahead horizon. As discussed in subsection 3.3.2, the benefit perceived by the FD participants is modelled as a function of the binary decision expressing whether the related demand activity is forgone or not, and the users' preferences and requirements associated with the operation of the FD participants are expressed solely in the form of constraints. Moreover, the hourly power demand of the inflexible demand participant is assumed constant and the benefit of which is not considered in this chapter.

This market clearing optimization is formulated as follows:

$$\max_{\{\xi_i, \forall i, \psi_j, \forall j\}} \sum_j B_j(\psi_j) - \sum_i C_i(\xi_i) \qquad (3.25)$$

subject to:

$$D_t + \sum_j d_{j,t} - \sum_i g_{i,t} = 0, \quad \forall t \qquad (3.26)$$

$$\xi_i \in G_i, \quad \forall i \qquad (3.27)$$

$$\psi_j \in D_j, \quad \forall j \qquad (3.28)$$

This problem is subject to the system-level constraint (3.26) that represents the coordination of the hourly demand-supply balance, and local-level constraints (3.27)—(3.28) associated with the operational characteristics of generation and FD participants (as comprehensively described in subsection 3.3.1 and subsection 3.3.2).

The values of the decision variables at the optimal solution constitute the market clearing schedules for generation and FD participants, denoted as $\xi_i^* \equiv [g_i^*, u_i^*]$; $\forall i \in I$, $\psi_j^* \equiv [d_j^*, v_j^*, w_j^*]$; $\forall j \in J^c$ and $\psi_j^* \equiv [d_j^*, v_j^*, z_j^*]$; $\forall j \in J^f$ respectively.

A continuous version of the problem (3.25)—(3.28) is solved subsequently with the binary variables set equal to their optimal values. The Lagrangian multipliers associated with constraint (3.26) constitute the centralized electricity price λ^*. This procedure is followed as Lagrangian multipliers cannot be directly calculated from the solution of problems in the presence of binary variables[34],[37].

Under the centralized market clearing prices and operating schedule, the centralized profit/utility of a generator i and an FD j are respectively expressed by:

$$pro_i(\lambda^*, \xi_i^*) = (\lambda^*)' g_i^* - C_i(\xi_i^*) \qquad (3.29)$$

$$uti_j(\lambda^*, \psi_j^*) = B_j(\psi_j^*) - (\lambda^*)' d_j^* \qquad (3.30)$$

3.4 Surplus Sub-Optimality Effects and Their Relation to Participants' Non-Convexities

3.4.1 Surplus Sub-Optimality

In a competitive market, generation and FD participants act as self-interested surplus-maximizing entities, given the prevailing electricity prices and subject to their operating constraints sets. These self-scheduling optimization problems for a generation i and a FD j are expressed by equation (3.31) and equation (3.32), and the resulting optimal schedules by ξ_i^s and ψ_j^s respectively.

$$\max_{\xi_i \in G_i}(pro_i \equiv (\boldsymbol{\lambda})' \boldsymbol{g}_i - C_i(\boldsymbol{\xi}_i)) \tag{3.31}$$

$$\max_{\psi_j \in D_j}(uti_j \equiv B_j(\boldsymbol{\psi}_j) - (\boldsymbol{\lambda})' \boldsymbol{d}_j) \tag{3.32}$$

As mentioned in Section 3.1, marginal pricing scheme cannot align the system's goal of centralized social welfare maximization and the participants' goals of individual surplus maximization. The rationale behind this lies in the fact that the marginal prices are calculated as the Lagrangian multipliers associated with demand-supply balance constraint (3.26), which means that the prices only reflect the level of variable generation costs, but do not encapsulate any non-variable costs (e.g. fixed, start-up, and shut-down costs) nor any non-convexities of generation. Therefore, some generators might be required to operate with a negative profit, as the marginal prices might not be sufficiently high to recover their fixed costs. Furthermore, some generation units might be required to operate with positive yet lower profit than the one they would obtain by self-scheduling given the marginal prices.

In other words, generators' self-schedules (given the posted marginal prices) are not generally consistent with the market clearing schedules. In cases of such inconsistencies, the latter schedules generally entail lower surpluses than the former, with this difference termed as profit loss, and this effect is termed profit sub-optimality. In the long run, these effects are undesirable since the operation of generation companies with profit sub-optimality and especially with negative profit will drive them either temporarily or permanently, out of business. Retiring a plant before the end of its economic lifetime or, more generally, before it can generate the expected return on investment renders it a stranded asset. Furthermore, such effects will also discourage necessary generation investments from other companies, endangering the security of the system.

Analogously, as a result of the non-convexities, the FD participants could incur *utility loss* given the marginal prices. The surplus loss for a generator i, an FD j and the total surplus loss are respectively expressed by:

$$\Delta loss_i^g \equiv pro_i(\boldsymbol{\lambda}^*, \boldsymbol{\xi}_i^s) - pro_i(\boldsymbol{\lambda}^*, \boldsymbol{\xi}_i^*) \geqslant 0 \tag{3.33}$$

$$\Delta loss_j^d \equiv uti_j(\boldsymbol{\lambda}^*, \boldsymbol{\psi}_j^s) - uti_j(\boldsymbol{\lambda}^*, \boldsymbol{\psi}_j^*) \geqslant 0 \tag{3.34}$$

$$TotalLoss = \sum_i \Delta loss_i^g + \sum_j \Delta loss_j^d \geqslant 0 \tag{3.35}$$

3.4.2 Generation Non-Convexities and Impacts on Surplus Optimality

On the basis of the generation operation model (subsection 3.3.1), in this subsection we examine the two representative generation non-convexities (Appendix A) associated with respectively i) fixed costs and ii) minimum stable generation (MSG) constraints coupled with the binary (on/off) commitment decisions, and illustrate their relationship to dispatch inconsistency and profit sub-optimality effects[34-45].

As discussed in subsection 3.3.1, regarding i), the generation cost functions include a term (fixed cost) that represents the cost components incurred whenever the generator is online, and irrespectively of its output level. Thus the cost function exhibits a discontinuity at the point of zero generation: when the generation becomes marginally higher than zero (which means that a generator is switched on), the cost "jumps" from zero to the value of the fixed cost of the committed generator. Due to this discontinuity, the generation cost functions are non-convex (Appendix A). The relation of this non-convexity with inconsistency and profit sub-optimality effects is demonstrated through a single-period problem, where the market includes a) a generator i with cost function $C_i(g_{i,t}) = 10 g_{i,t} + 50 u_{i,t}$, and with $g_i^{\min} = 0$ and $g_i^{\max} = 20$ MW, and b) inflexible demand with $D_t = 15$ MW. Centralized market clearing involves dispatching this generator to meet the inflexible demand, and thus yields $g_{i,t}^* = D_t = 15$ MW, $u_{i,t}^* = 1$, and $\lambda_t^* = 10$ £/MWh. Under the centrally imposed schedule, generator i is operating at a negative profit of $-£50$. This is due to the fact that marginal price of 10 £/MWh is not high enough to recover the fixed cost of £50. Consequently, given this price, generator i would prefer to turn off in order to avoid this profit deficit. Thus, generator i entails a profit loss of £50 under the centralized solution.

Regarding ii), the operation of generation units is usually subject to minimum stable generation constraints to ensure the reliable and stable operation of the unit. The combination of the unit commitment decisions and the MSG constraints results in a non-convex operation domain (Appendix A), since generator's output can take the values of 0 and $g_i^{\min}$ but not any value in-between. A single-period problem is considered here to illustrate the consequences of this non-convexity. The market includes a) two generators: generator 1 with cost function $C_1(g_{1,t}) = 12 g_{1,t}$, and power limits of $g_1^{\min} = 0$ MW and $g_1^{\max} = 10$ MW; generator 2 with cost function $C_2(g_{2,t}) = 10 g_{2,t}$, and power limits of $g_2^{\min} = 10$ MW and $g_2^{\max} = 20$ MW, and b) inflexible demand with $D_t = 5$ MW. Although generator 2 is cheaper than generator 1, the minimum output restriction of generator 2 is higher than the demand, hence the centralized market clearing solution involves $g_{1,t}^* = D_t = 5\,\text{MW}$, $u_{1,t}^* = 1$ and $\lambda_t^* = 12$ £/MWh. While generator 2 is off in the centralized scheduling, it would want to output at its maximum power limit in the self-scheduling at the price of 12 £/MWh in order to fulfill a profit of £40. As a result, generator 2 incurs a profit loss of £40 under the centralized solution.

3.4.3 Flexible Demand Non-Convexities and Impacts on Surplus Optimality

In line with the previous subsection, FD participants are examined from the same perspective. On the basics of the FD models proposed in subsection 3.3.2, three different types of non-convexities are recognized.

The first FD non-convexity (Appendix A) is associated with the potential to forgo demand activities and is mathematically captured by the binary decision variable v_j. The relation of this non-convexity with inconsistency and surplus sub-optimality effects is demonstrated through a single-period problem, where the market includes a) an FCFD j with $T_j^{\text{dur}} = 1$, $d_{j,1}^{\text{cyc}} = 10\,\text{MW}$, $B_j^0 = £150$ and only able to forgo its activity, and b) a generator i with a cost function $C_i(g_{i,t}) = g_{i,t}^2$ and without any non-convex characteristics. Centralized market clearing involves carrying out the FCFD activity, since B_j^0 is higher than the generation cost incurred to satisfy the demand of this activity (£100), and thus yields $g_{i,t}^* = d_{j,t}^* = 10\,\text{MW}$ and $\lambda_t^* = 20$ £/MWh. However, given this price, the FCFD would choose to forgo its activity, since its payments ($\lambda_t^* \times d_{j,t}^* = £200$) are higher than its benefits ($B_j = B_j^0 = £150$); therefore, under the centralized solution, the FCFD incurs a utility loss of £50.

Let us now neglect the potential to forgo the activity and only consider the ability to redistribute the activity across time. FCFDs still have a non-convex operating constraints' set (Appendix A), since their demand $d_{j,t}$ during their scheduling window $[t_j^{\text{start}}, t_j^{\text{end}}]$ can only take a set of discrete values, including 0 and the fixed power requirement $d_{j,\tau}^{\text{cyc}}$ of each step of their cycle.

A two-period problem is considered here, where the market includes a) an FCFD j with $T_j^{\text{dur}} = 1$, $d_{j,1}^{\text{cyc}} = 12\,\text{MW}$, $t_j^{\text{start}} = 1$, and $t_j^{\text{end}} = 2$ i.e. the FCFD can carry out its activity at either $t = 1$ or $t = 2$, b) inflexible demands with $D_1 = 10\,\text{MW}$ and $D_2 = 20\,\text{MW}$ and c) the same generator i with the previous example. Centralized market clearing schedules the FCFD activity at $t = 1$ in order to flatten as much as possible the total demand profile and minimize the total generation costs, and thus yields $d_{j,1}^* = 12\,\text{MW}$, $d_{j,2}^* = 0$, $\lambda_1^* =$

44 £/MWh and $\lambda_2^* = 40$ £/MWh. However, given these prices, the FCFD would choose to carry out its activity at $t=2$ since this period exhibits a lower price; therefore, under the centralized solution, the FCFD incurs a utility loss of £48.

CCFDs with the redistributing ability also exhibit a non-convex operating constraints set (Appendix A), since their power demand can take the values $d_{j,t} = 0$ and $d_{j,t} = d_{j,t}^{\min}$ but not any value in the range $(0, d_{j,t}^{\min})$. A two-period problem is also considered here, where the market includes a) a CCFD j with $E_j = 12$ MWh, $d_{j,1}^{\min} = d_{j,2}^{\min} = 5$ MW, $d_{j,1}^{\max} = d_{j,2}^{\max} = 15$ MW, b) the same inflexible demands with the previous example and c) the same generator i with the previous examples. Centralized market clearing schedules the CCFD activity entirely at $t=1$ in order to flatten as much as possible the total demand profile and minimize the total generation costs, and thus yields $d_{j,1}^* = 12$ MW, $d_{j,2}^* = 0$, $\lambda_1^* = 44$ £/MWh, and $\lambda_2^* = 40$ £/MWh. However, given these prices the CCFD would choose to be scheduled entirely at $t=2$ since this period exhibits a lower price; therefore, under the centralized solution, the CCFD incurs a utility loss of £48. If the same problem is considered with $d_{j,1}^{\min} = d_{j,2}^{\min} = 0$, the optimal centralized solution flattens completely the total demand profile with $d_{j,1}^* = 11$ MW, $d_{j,2}^* = 1$ MW and $\lambda_1^* = \lambda_2^* = 42$ £/MWh. Given these prices, any feasible solution of the CCFD's self-scheduling problem — including its above centralized schedule — is an optimal one, and therefore the CCFD does not incur utility loss.

As mentioned in subsection 3.3.2, the FD model can incorporate alternative benefit functions to capture more accurately the preferences of some consumers, and employ more constraints to capture practical aspects of some flexible load operation. However, since the modelling of more complex benefit functions and the additional operating constraints do not introduce additional binary variables (other than v_j, w_j and z_j), they do not introduce additional FD non-convexities.

As mentioned in subsection 3.3.2, FD participants generally correspond to large industrial/commercial consumers, or FD aggregators, representing a large number of smaller flexible residential/commercial consumers. In the examples examined in this subsection, we considered a small number of large FDs. Both FD types exhibit a non-convex operating constraints set, and as a result the optimal price response of the FDs involves lumpy rather than smooth demand shifts towards the lowest-priced hours of the market horizon. This impact will be exacerbated as the relative number, consumption size and flexibility extent of flexible loads with respect to the temporal variation of inflexible demands in the system increases, and their homogeneity enhances. Consequently, if we consider a large number of small FDs, although the lumpiness will still be present in the aggregated load profile; due to the enhanced load diversity, the lumpy demand shifts will be smoothed out to some extent, and the schedule inconsistency effects will be less significant.

To sum up, the identified FDs' non-convexities are associated with their ability to forgo activities, as well as discrete and minimum power levels. It should be noted that the presented FD non-convexities are not identical with the respective non-convexities of the generation side, given their significant operational differences. More specifically, the non-convexity associated with the ability of FD to forgo demand activities and the non-convexity associated with discrete demand levels certainly do not apply to the generation side.

3.5 Generalized Uplifts Under Flexible Demand Participation

As stressed in Section 3.1 and Section 3.4, the competitive market equilibrium solution may be unattainable if linear uniform pricing mechanism is adopted. In order to adequately incentivize participants and eliminate any surplus sub-optimality and schedule inconsistency effects, a combination of *market-based payments* (payment derived from the market prices and are paid within the market) and *side payments* or *uplift payments* (charges not collected as part of the market-based payments and are paid outside the market) is generally adopted by the market operators[34-45]. The developed approaches can be broadly divided into two categories: *lump-sum uplifts* and *generalized uplifts*.

3.5.1 Lump-Sum Uplifts

Regarding the lump-sum uplift approach, the market operator retains uniform pricing, and assigns uplift payments to participants (only if they endure surplus sub-optimality) that compensates exactly their respective surplus losses calculated at the prevailing market prices. In our case, under marginal prices λ^*, generator i and FD j experiencing surplus sub-optimality receive uplift payments of U_i^{g*} and U_j^{d*}, respectively:

$$U_i^{g*} = \Delta\, loss_i^g = pro_i(\lambda^*, \xi_i^s) - pro_i(\lambda^*, \xi_i^*) \geqslant 0 \qquad (3.36)$$

$$U_j^{d*} = \Delta\, loss_j^d = uti_j(\lambda^*, \psi_j^s) - uti_j(\lambda^*, \psi_j^*) \geqslant 0 \qquad (3.37)$$

Due to the fact that the surplus loss in market with non-convexities is always non-negative for any uniform pricing schemes[37-38], [44], the required total uplift payment is also always non-negative. However, uniform prices differ themselves regarding the amount of the requisite uplift payments to support the competitive equilibrium[37-38]. As quantitatively demonstrated in subsection 3.7.4 and Appendix B, marginal pricing results in high overall lump-sum uplift payment (thus surplus loss) as opposite to convex hull pricing which corresponds to the minimum uplift payment. Given that uplift payments are generally not desirable as they are considered as an *out-of-market intervention*, convex hull pricing is more favorable in this regard[37-40].

Upon receiving the uplift payments, participants' surpluses are made equal to their maximum level, given that conservation of monetary flow within the market should be satisfied, this means that the total compensation for the surplus loss is entirely collected from the (inflexible) demand side of the market, which is thus treated inequitably. In our case, a lump-sum monetary transfer is made from the inflexible demand side to the generation and FD participants, through a negative uplift $U^{\inf *}$ balancing the positive uplifts received by the rest of the participants:

$$U^{\inf *} \equiv -\sum_i U_i^{g*} - \sum_j U_j^{d*} \tag{3.38}$$

3.5.2 Generalized Uplifts

To address the aforementioned fairness issue, references [43]—[45] propose the implementation of generalized uplifts that modifies a generator's profit function by adding an uplift function. By contrast the lump-sum uplift approach, all market participants not operating at a loss without uplifts will contribute to compensate the overall surplus sub-optimality. By invoking the rule of sharing surplus loss compensation among participants, the market operator not only reduces the bills of consumers, but also provides appropriate impetuses to the generators not trying to artificially cause profit sub-optimality through offering strategies[45].

The *generalized uplift functions U* constitute additional revenues ($U > 0$) or payments ($U < 0$) for the generators. $U > 0$ implies generators are operating at a loss without uplifts so that they are compensated by the positive uplift payments. $U < 0$ suggests that generators do not experience surplus loss without uplifts, thus they provide a subsidy (through a negative uplift payment) to generators that need the compensation. It should be noted that negative uplifts are introduced in both lump-sum and generalized uplift approaches; the difference is that the former entails a negative uplift only for the demand side, while the latter entails negative uplifts for some of the other market participants in order to yield a more equitable distribution of the total surplus loss compensation.

In reference [45], the generalized uplift function U_i^g (3.39) applying to generator i includes a set of adjustable generator-specific parameters $\Delta \pi_i^g = [\Delta \alpha_i^g, \Delta c_i^{on}, \Delta c_i^{off}]$ associated with the power output, the "on" commitment status and the "off" commitment status of generator i respectively.

$$U_i^g(\xi_i, \Delta \pi_i^g) = \sum_t [\Delta \alpha_{i,t}^g g_{i,t} + \Delta c_{i,t}^{on} u_{i,t} + \Delta c_{i,t}^{off}(1-u_{i,t})] \quad (3.39)$$

These parameters provide sufficient degrees of freedom for the market operator to achieve the competitive equilibrium while supporting the quantities in an economic dispatch. More specifically, all three parameters adjust the self-scheduling surplus of generator i, $\Delta \alpha_{i,t}^g$ enforces generator i to produce at the optimal output level according to the centralized solution, while $\Delta c_{i,t}^{on}$ and $\Delta c_{i,t}^{off}$ incentivize generator i to follows the optimal unit commitment statuses determined by the market clearing.

In the same vein, we propose a generalized uplift function U_j^d (3.40) applying to FD j. It includes a set of adjustable FD-specific parameters $\Delta \pi_j^d = [\Delta \alpha_j^d, \Delta \gamma_j^d]$ associated with the power input and forgoing the activity of FD j respectively.

$$U_j^d(\psi_j, \Delta \pi_j^d) = \sum_t \Delta \alpha_{j,t}^d d_{j,t} + \Delta \gamma_j^d (1-v_j) \quad (3.40)$$

The rationale behind parameter $\Delta \alpha_{j,t}^d$ is to adjust its self-scheduling surplus if its demand activity is carried out under centralized market clearing, so that FD j is incentivized to carry out its activity at the optimal periods according to the centralized solution. $\Delta \gamma_j^d$ is required to adjust the self-scheduling surplus of FD j if its demand activity is forgone under centralized market clearing, so that FD j is incentivized to forgone its activity as well in the self-scheduling. The detailed demonstration on how these two parameters deal with surplus sub-optimality effects is provided in subsection 3.7.3.

Given the above uplift functions, the *augmented self-scheduling* optimization problems involve the maximization of the augmented profit of generator i (3.41) and the augmented utility of FD j (3.42).

$$pro_i(\lambda^N, \xi_i, \Delta \pi_i^g) \equiv (\lambda^N)' g_i - C_i(\xi_i) + U_i^g(\xi_i, \Delta \pi_i^g) \quad (3.41)$$

$$uti_j(\lambda^N, \psi_j, \Delta \pi_j^d) \equiv B_j(\psi_j) - (\lambda^N)' d_j + U_j^d(\psi_j, \Delta \pi_j^d) \quad (3.42)$$

3.5.3 Formulation of Minimum Discrimination Problem

Although the generalized uplift approach yields more equitable distribution of the total surplus loss compensation, it introduces price discrimination among the market participants that cannot be easily justified and may be considered non-transparent[34],[37-38]. To this end, the calculation of these differentiated parameters in reference [44] is carried out through an optimization problem minimizing the extent of discrimination introduced (thus denoted as the *minimum discrimination problem* in the rest of this chapter). In this subsection, the mechanism developed in references [44]—[45] is extended to account for the FD non-convexities (subsection 3.4.3) by incorporating the proposed FD generalized uplift functions (subsection 3.5.2).

The mathematical formulation of the proposed minimum discrimination problem is presented as follows:

Objective function:

The objective of minimizing the discrimination introduced by the differentiated pricing terms is expressed by the minimization of the square norm of uplift parameters [44]:

$$\min_{(\Delta \pi_i^g, \forall i, \Delta \pi_j^d, \forall j)} \sum_i \|\Delta \pi_i^g\|^2 + \sum_j \|\Delta \pi_j^d\|^2 \tag{3.43}$$

The rationale behind this objective function lies in minimizing the extent of discrimination introduced by minimizing the sum of the square norms of these parameters. Given that the original motivation behind the employment of the generalized uplift approach is to support the competitive equilibrium while enforcing a more equitable distribution of the total surplus loss, the goal of this approach is to achieve such equitable distribution with the minimum extent of discrimination introduced by the participant-specific parameters.

Constraints:

1) The solution of the centralized market clearing problem is identical to the solution of the augmented self-scheduling problems for all market participants.

$$\xi_i^* = \xi_i^a \equiv \arg \max_{\xi_i \in G_i} pro_i(\lambda^N, \xi_i, \Delta \pi_i^g), \ \forall i \tag{3.44}$$

$$\psi_j^* = \psi_j^a \equiv \arg \max_{\psi_j \in D_j} uti_j(\lambda^N, \psi_j, \Delta \pi_j^d), \ \forall j \tag{3.45}$$

2) Conservation of monetary flow within the market should be satisfied, meaning that the market operator is revenue neutral. This imposes that the total revenues collected by

the consumers are equal to the total payments paid to the generators, including the uplifts (equation (3.46)). The combination of equation (3.46) with the demand-supply balance (3.26) implies that the sum of uplifts should be zero.

$$\sum_j [(\lambda^N)' d_j^* - U_j^d(\psi_j^*, \Delta \pi_j^d)] + (\lambda^N)'D = \sum_i [(\lambda^N)' g_i^* + U_i^g(\xi_i^*, \Delta \pi_i^g)] \quad (3.46)$$

$$\sum_i U_i^g(\xi_i^*, \Delta \pi_i^g) + \sum_j U_j^d(\psi_j^*, \Delta \pi_j^d) = 0 \quad (3.47)$$

3) The difference between a participant's surplus under augmented self-scheduling and self-scheduling without uplifts determines the contribution of this participant to the compensation for the total surplus loss[44] and is expressed by (3.48), (3.49), and (3.50) for generator i, FD j, and the inflexible demand① respectively. This contribution should not be negative to ensure that participants do not derive exceptional surplus from the uplifts and the new electricity prices; in other words, under augmented self-scheduling, each participant should derive at most their maximum surplus under self-scheduling without uplifts[44].

$$\Delta cont_i^g(\lambda^N, \Delta \pi_i^g) \equiv pro_i(\lambda^*, \xi_i^s, 0) - pro_i(\lambda^N, \xi_i^a, \Delta \pi_i^g) \geqslant 0 \quad (3.48)$$

$$\Delta cont_j^d(\lambda^N, \Delta \pi_j^d) \equiv uti_j(\lambda^*, \psi_j^s, 0) - uti_j(\lambda^N, \psi_j^a, \Delta \pi_j^d) \geqslant 0 \quad (3.49)$$

$$\Delta cont^{\text{inf}}(\lambda^N) \equiv (\lambda^N)'D - (\lambda^*)'D \geqslant 0 \quad (3.50)$$

The combination of (3.26), (3.33)—(3.35), (3.44)—(3.45), and (3.47)—(3.50) yields (3.51), which expresses the fact that the total surplus loss is equal to the total compensation contribution by all participants.

$$Loss^{\text{Total}} = \Delta cont^{\text{inf}}(\lambda^N) + \sum_i \Delta cont_i^g(\lambda^N, \Delta \pi_i^g) + \sum_j \Delta cont_j^d(\lambda^N, \Delta \pi_j^d) \quad (3.51)$$

The market arrangements should generally facilitate an equitable distribution of the compensation for the total surplus loss among the participants. In this context, authors in reference [44] proposed two market rules: a) the total compensation is divided equally among generators and (inflexible) demand, b) the ratio between profit under augmented self-scheduling and self-scheduling without uplifts is set equal for all generators. Given that an equitable compensation distribution between generators, FDs,

---

① Given that the benefit of inflexible demand is not considered in this chapter, strictly speaking, its contribution is not expressed by the difference between its surplus under augmented self-scheduling and self-scheduling without uplifts, but by the difference between its respective payments.

and inflexible demand cannot be determined unambiguously, we propose an extended version of the latter market rule, where the ratio between surplus under augmented self-scheduling and self-scheduling without uplifts is set equal to a common value R ($0 < R < 1$) for all generators, FDs, and inflexible demand[①].

$$pro_i(\pmb{\lambda}^N, \pmb{\xi}_i^a, \Delta\pmb{\pi}_i^g) = pro_i(\pmb{\lambda}^*, \pmb{\xi}_i^s, \pmb{0}) \times R, \ \forall i \tag{3.52}$$

$$uti_j(\pmb{\lambda}^N, \pmb{\psi}_j^a, \Delta\pmb{\pi}_j^d) = uti_j(\pmb{\lambda}^*, \pmb{\psi}_j^s, \pmb{0}) \times R, \ \forall j \tag{3.53}$$

$$(\pmb{\lambda}^*)'\pmb{D} = (\pmb{\lambda}^N)'\pmb{D} \times R \tag{3.54}$$

Substituting (3.52)—(3.54) into (3.51) yields:

$$TotalLoss = \overbrace{\left[\underbrace{\sum_{i \in I} pro_i(\pmb{\lambda}^*, \pmb{\xi}_i^s, \pmb{0})}_{A} + \underbrace{\sum_{j \in J} uti_j(\pmb{\lambda}^*, \pmb{\psi}_j^s, \pmb{0})}_{B}\right] \times (1-R)} + \underbrace{(\pmb{\lambda}^*)'\pmb{D}}_{C} \times \left(\frac{1}{R} - 1\right)$$

$$\tag{3.55}$$

Along with the condition $0 < R < 1$, R is calculated as:

$$R = \frac{-(A+C-B) + \sqrt{(A+C-B)^2 + 4BC}}{2B} \tag{3.56}$$

Equation (3.54), along with the assumption that the electricity price is uniformly increased across all periods of the market horizon[44], fixes the new electricity prices according to (3.57).

$$\lambda_t^N = \lambda_t^* + \frac{\left(\frac{1}{R} - 1\right)\sum_t \lambda_t^* D_t}{\sum_t D_t}, \ \forall t \tag{3.57}$$

The new values of the electricity prices do not possess a strict physical meaning but are adjusted to ensure that the inflexible demand contributes at a desired level (according to equation (3.50)) to the total surplus loss compensation. This desired level of compensation can be achieved with many different combinations of the new hourly electricity prices λ_t^N. The assumption of uniform increase of the electricity price across all periods of the market horizon is made in both reference [44] and our approach in order to reduce the number of variables and thus the computational complexity of the minimum discrimination problem. Specifically, with this assumption, the new electricity

---

① In the same line as the previous footnote, strictly speaking, the ratio R does not relate the surplus under augmented self-scheduling and self-scheduling without uplifts in the case of inflexible demand, but its respective payments.

prices are fixed according to (3.57), thus the decision variables of the minimum discrimination problem include only the participant-specific uplift parameters.

3.5.4 Solution Techniques of the Minimum Discrimination Problem

In order to solve the minimum discrimination problem, the optimal solutions ξ_i^a, $\forall i$ and ψ_j^a, $\forall j$ of the augmented self-scheduling problems need to be analytically expressed in terms of the uplift parameters and new electricity prices, so as to enforce the equal schedule conditions (3.44)—(3.45). As discussed in reference [45], such analytical derivations are impractical for multi-period market clearing problems accounting for participants' time-coupling characteristics. In order to address this challenge, authors in reference [45] proposed an iterative cutting-plane algorithm for the solution of the minimum discrimination problem, and proved its convergence and optimality. This algorithm iteratively restricts the feasible set of uplift parameters and new electricity prices, through the sequential generation of profit cutting planes, in order to impose indirectly surplus optimality conditions. In this section, this algorithm is extended to compute the uplift parameters of both generation and FD participants, and its convergence performance is improved through the employment of an additional termination criterion. The algorithm includes the following steps:

(**Step 1**) The centralized market clearing problem (3.25)—(3.28) is solved to obtain ξ_i^*, $\forall i$, ψ_j^*, $\forall j$, and λ^*. The iteration counter is set to zero $r = 0$, as well as the initial values of the uplift parameters $(\Delta \pi_i^g)^{[0]} = \mathbf{0}$, $\forall i$ and $(\Delta \pi_j^d)^{[0]} = \mathbf{0}$, $\forall j$. The sets of generators' profit and FDs' utility cutting planes are initialized to the entire space of uplift parameters $\Xi_i^{[0]} = \mathbb{R}^{3N_T}$, $\forall i$ and $\Psi_j^{[0]} = \mathbb{R}^{2N_T+1}$, $\forall j$. The new electricity prices λ^N are fixed by the equitable compensation distribution rule (3.57), and therefore constitute fixed inputs to the remaining steps of the algorithm.

(**Step 2**) The augmented self-schedule of each market participant given the latest value of the uplift parameters and the new electricity prices, is calculated:

$$(\xi_i^a)^{[r+1]} = \arg\max_{\xi_i \in G_i} pro_i(\lambda^N, \xi_i, (\Delta \pi_i^g)^{[r]}), \forall i \qquad (3.58)$$

$$(\psi_j^a)^{[r+1]} = \arg\max_{\psi_j \in D_j} uti_j(\lambda^N, \psi_j, (\Delta \pi_j^d)^{[r]}), \forall j \qquad (3.59)$$

(**Step 3**) According to reference [45], if the augmented self-schedule of a generator i or an FD j obtained in (Step 2) is different from the respective centralized schedule, a new constraint cut is added to the respective cutting plane set. An additional criterion for the generation of new constraint cuts is employed, in order to account for the fact that problems (3.58)—(3.59) do not necessarily have unique solutions, due to their non-convex, mixed-integer nature. This criterion involves the comparison between the surpluses under the augmented self-schedules obtained in (Step 2) and the respective centralized schedules. If these two surpluses are identical for a participant, then the latter does not witness surplus sub-optimality and therefore a new cut is not added in its cutting plane set, even if its augmented self-schedule and centralized schedule are not identical.

The employment of this additional criterion improves the convergence performance of the algorithm. This is because it avoids the execution of redundant iterations in the case some of the problems (3.58)—(3.59) have multiple equivalent solutions including the respective centralized schedule, but the latter is not selected by the computational algorithm solving (3.58)—(3.59). To sum-up, if for a generator i or an FD j respectively:

$$pro_i(\lambda^N, (\xi_i^a)^{[r+1]}, (\Delta \pi_i^g)^{[r]}) \neq pro_i(\lambda^N, \xi_i^*, (\Delta \pi_i^g)^{[r]}) \text{ AND } (\xi_i^a)^{[r+1]} \neq \xi_i^*$$
(3.60)

$$uti_j(\lambda^N, (\psi_j^a)^{[r+1]}, (\Delta \pi_j^d)^{[r]}) \neq uti_j(\lambda^N, \psi_j^*, (\Delta \pi_j^d)^{[r]}) \text{ AND } (\psi_j^a)^{[r+1]} \neq \psi_j^*$$
(3.61)

then respectively (where ε^g and ε^d are small, strictly positive parameters[45]):

$$\Xi_i^{[r+1]} = \Xi_i^{[r]} \cap \{\Delta \pi_i^g \in \mathbb{R}^{3T} : pro_i(\lambda^N, (\xi_i^a)^{[r+1]}, \Delta \pi_i^g) \leqslant pro_i(\lambda^N, \xi_i^*, \Delta \pi_i^g) - \varepsilon^g\}$$
(3.62)

$$\Psi_j^{[r+1]} = \Psi_j^{[r]} \cap \{\Delta \pi_j^d \in \mathbb{R}^{T+1} : uti_j(\lambda^N, (\psi_j^a)^{[r+1]}, \Delta \pi_j^d) \leqslant uti_j(\lambda^N, \psi_j^*, \Delta \pi_j^d) - \varepsilon^d\}$$
(3.63)

else respectively:

$$\Xi_i^{[r+1]} = \Xi_i^{[r]}$$
(3.64)

$$\Psi_j^{[r+1]} = \Psi_j^{[r]}$$
(3.65)

(**Step 4**) If conditions (3.64)—(3.65) are satisfied for every participant, the algorithm terminates. If not, a new set of trial uplift parameters $(\Delta \pi_i^g)^{[r+1]}$ and $(\Delta \pi_j^d)^{[r+1]}$ is calculated through the solution of problem (3.43) subject to the market operator's revenue neutrality constraints(3.47), the equitable compensation distribution constraints (3.52)—(3.54) and constraint (3.57), and the constraints of the cutting plane sets (constraints (3.66)—(3.67)). The iteration counter is then incremented by 1 and the algorithm returns to (Step 2).

$$\Delta \pi_i^g \in \Xi_i^{[r+1]}, \forall i \tag{3.66}$$

$$\Delta \pi_j^d \in \Psi_j^{[r+1]}, \forall j \tag{3.67}$$

3.6 Convex Hull Pricing Under Flexible Demand Participation

3.6.1 Concept of Convex Hull Pricing

In this subsection, we build an intuition regarding the concept of convex hull pricing from a graphical perspective. Consider a general unit commitment problem (see the generic optimization model of this problem in Appendix B) where the market operator dispatches generators to supply demand by solving a total generation cost minimization problem. The *value function*[37-39], defined as the optimal total cost as a function of the load level for the unit commitment problem, is constructed. Due to the non-convex operational characteristics of the generation participants (as identified in subsection 3.4.2), this value function exhibits non-continuities and non-convexities, but is differentiable almost everywhere (except at the discontinuities). Under the marginal pricing mechanism, market prices are derived as the sub-gradient (Appendix B) of the value function at a specific load level. Under the convex hull pricing mechanism, one needs to construct the *convexified value function* defined as the greatest convex function (namely the *convex hull* or *convex envelope*) that is nowhere greater than the value function. Subsequently, the convex hull prices are defined as the sub-gradient (Appendix B) of the convex hull of the value function (rather than the value function itself)[37-39].

Consider the following single-period example, where the market operator solves a unit commitment problem to meet load d. Generator 2 is characterized by a higher variable cost ($l_2^G > l_1^G$), but a much lower fixed cost ($fc_2^G \ll fc_1^G$), and a lower minimum stable and maximum generation limits ($g_2^{min} < g_1^{min}$ and $g_2^{max} < g_1^{max}$). The optimal generation dispatches and unit commitment statuses for different load levels are presented in Table 3.1, and these results are explored as follows:

- $d = 0$

Neither of the generators are dispatched.

- $d = [0, g_2^{min}]$

Since the load level is less than neither generator's MSG level, the unit commitment problem is infeasible in this range of load.

- $d \in \left[g_2^{\min}, \dfrac{fc_2^G - fc_1^G}{l_2^G - l_1^G} \right)$

Generator 2 is dispatched to meet demand, since generator 2 has a smaller MSG level and generator 1 exhibits a much higher fixed cost (although it exhibits a lower variable cost, overall it is cheaper to dispatch generator 2).

- $d \in \left[\dfrac{fc_2^G - fc_1^G}{l_2^G - l_1^G}, g_1^{\max} \right)$

As load level increases beyond $\dfrac{fc_2^G - fc_1^G}{l_2^G - l_1^G}$, dispatching generator 1 instead to meet demand corresponds to a more cost-effective solution.

- $d \in [g_1^{\max}, g_1^{\max} + g_2^{\min})$

As load level increases beyond $g_1^{\max}$, it necessitates the commitment of generator 2. The minimum-cost solution involves generator 2 outputting at $g_2^{\min}$ (due to its MSG limit) while the rest of load is supplied by generator 1.

- $d \in [g_1^{\max} + g_2^{\min}, g_1^{\max} + g_2^{\max}]$

As load level increases beyond $g_1^{\max} + g_2^{\min}$, the least-cost solution involves scheduling generator 1 at its full capacity while the rest of load is satisfied by generator 2.

The value function, its convex hull function and their respective sub-gradients for the examined example can be visualized in Figure 3.3. The value function (the solid curve) is constructed on a discontinuous demand set as a consequence of the MSG constrains (as analysed above). It exhibits discontinuities or "jumps" as a result of the changes in commitment statues in variation of load level (Table 3.1). The convex hull is the closest

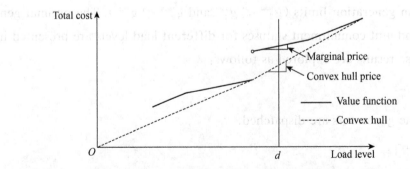

Figure 3.3 Value function and its convex hull, marginal price and convex hull price

convex approximation from below (the dash curve). It should be noted that both the value function and its convex hull include the origin point. The slopes of the value function and its convex hull at load level d determine the marginal price and convex hull price respectively in supplying that load, as highlighted in Figure 3.3.

Table 3.1 Generation dispatch and unit commitment status for different load level

d	0	$\left[g_2^{\min}, \dfrac{fc_2^G - fc_1^G}{l_2^G - l_1^G}\right)$	$\left[\dfrac{fc_2^G - fc_1^G}{l_2^G - l_1^G}, g_1^{\max}\right)$	$[g_1^{\max}, g_1^{\max} + g_2^{\min})$	$[g_1^{\max} + g_2^{\min}, g_1^{\max} + g_2^{\max}]$
u_1^*	0	0	1	1	1
g_1^*	0	0	d	$d - g_2^{\min}$	$g_1^{\max}$
u_2^*	0	1	0	1	1
g_2^*	0	d	0	$g_2^{\min}$	$d - g_1^{\max}$

3.6.2 Lagrangian Formulation of Convex Hull Pricing Problem

Although the graphical formulation of convex hull pricing are considered transparent for simple single-period problems, for realistic market problems that include multiple time periods, and incorporating participants' inter-temporal characteristics, the construction of the value function and its convex hull becomes non-trivial and computationally intensive (for example every point on value function requires the solution of the centralized market clearing problem (3.25)—(3.28)). To this end, references [37]—[39] provide an alternative, tractable solution technique by relating the convex hull prices to the solutions of the *Lagrangian dual problem*, and have proven the validity of the approach. According to references [37]—[39], the convex hull prices λ^{CH} identify uniform energy prices that minimize the total uplift payments (thus the total surplus loss), and they coincide with the Lagrangian multipliers optimizing the dual problem (3.68) of the market clearing problem (3.25)—(3.28) (the connection of the convex hull pricing and Lagrangian dual maximization is comprehensively derived in Appendix B).

$$\max_{\lambda} \varphi(\lambda) = \max_{\lambda} \min_{\{\xi_i \in G_i, \forall i, \psi_j \in D_j, \forall j\}} L \quad (3.68)$$

$$L = \sum_i C_i(\xi_i) - \sum_j B_j(\psi_j) + (\lambda)' \left(\sum_j d_j + D - \sum_i g_i\right) \quad (3.69)$$

where φ is the dual function and L is the Lagrangian function of the problem (3.25)—

(3.28).

Given that the Lagrangian function constitutes an additive combination of the individual participants' surpluses, the dual problem is decomposed into the independent surplus maximization sub-problems (3.31), $\forall i$ and (3.32), $\forall j$ coordinated iteratively by a λ update algorithm until the maximum φ is achieved. This mathematical decomposition can be interpolated as a two-level iterative market clearing mechanism. At the local level, participants solve independently their individual surplus maximization problems for the given values of the prices, and submit their optimal demand/generation responses to the market operator. At the global level, the latter updates the value of λ posted to the participants, according to their responses, in an effort to gradually maximize φ, and the respective price at the maximum dual φ^* is the convex hull price λ^{CH}. The *duality gap* (Appendix B), defined as the difference between the primal optimum f^* (optimal solution of the primal problem (3.25)—(3.28)) and the dual optimum φ^*, identifies the minimum surplus loss, or equivalently the minimum uplift payment under λ^{CH} [37-39].

According to references [3]—[6] after the convex hull prices λ^{CH} have been determined by the above dual optimization problem, generator i and FD j experiencing surplus sub-optimality receive lump-sum uplifts $U_i^{CH, g}$ and $U_j^{CH, d}$ respectively, exactly compensating their respective surplus loss:

$$U_i^{CH, g} \equiv pro_i(\lambda^{CH}, \xi_i^s) - pro_i(\lambda^{CH}, \xi_i^*) \geqslant 0 \quad (3.70)$$

$$U_j^{CH, d} \equiv uti_j(\lambda^{CH}, \psi_j^s) - uti_j(\lambda^{CH}, \psi_j^*) \geqslant 0 \quad (3.71)$$

This means that the final generators' and FDs' surpluses are made equal to their maximum surpluses. Given that conservation of monetary flow within the market should be satisfied, this means that the total compensation of surplus loss is entirely charged to the inflexible demand through a negative lump-sum uplift $U^{CH, inf}$ balancing the positive uplifts received by the rest of the participants:

$$U^{CH, inf} \equiv -\sum_i U_i^{CH, g} - \sum_j U_j^{CH, d} \quad (3.72)$$

3.6.3 Solution Techniques of the Lagrangian Dual Problem

As discussed in subsection 3.6.2, the Lagrangian multipliers' update process is mathematically translated to a solution process of the dual function maximization. Therefore, the market operator should employ effective Lagrangian multipliers' update algorithms that are efficient in reaching this value in relatively low number of iterations (i.e. exhibit favorable convergence properties) and with relatively low computational burden per iteration.

Based on the definition of φ (3.68) and the Lagrangian function L (3.69), it is deduced that φ constitutes an additive combination of the individual participants' demand and supply functions. Since these functions exhibit discontinuities due to their non-convex characteristics (subsection 3.4.2 and subsection 3.4.3), φ is non-differentiable at certain points. Consequently, optimization techniques for non-differentiable functions should be employed for the multipliers' update procedure[93-98].

Among these techniques, *sub-gradients methods*[93-95] are preferred in the majority of the relevant literatures due to their conceptual simplicity and lower required computational time per iteration. These methods update each multiplier proportionally to the respective relaxed system constraint violation $e_t^{[r]}$ at the latest iteration according to (3.73)—(3.74).

$$\lambda_t^{[r+1]} = \lambda_t^{[r]} + s^{[r]} e_t^{[r]}, \forall t \qquad (3.73)$$

where:

$$e_t^{[r]} = D_t + \sum_{j \in J} d_{j,t}^{[r]} - \sum_{i \in I} g_{i,t}^{[r]} \qquad (3.74)$$

where $s^{[r]}$ represents the step length at iteration r. Although sub-gradient methods are not ascending methods — meaning that an increase in the value of the dual function after every iteration is not guaranteed — the distance between the latest values of the Lagrangian multipliers and the respective values at the dual function maximum gradually decreases, and the convergence of this algorithm after a finite number of iterations is guaranteed, if the step size satisfies the following conditions[94]:

$$\lim_{r \to \infty} s^{[r]} = 0 \qquad (3.75)$$

$$\sum_{r=1}^{\infty} s^{[r]} \to \infty \qquad (3.76)$$

The step size update rule satisfying these conditions and most commonly used in the relevant literature [93]—[95] updates the step size according to (3.77), where α and β are heuristically determined positive scalars and constitute the parameters of this method.

$$s^{[r]} = \frac{1}{(\alpha + \beta r)\|e^{[r]}\|} \qquad (3.77)$$

A suitable adjustment of these heuristic parameters has a significant impact on the performance of the multipliers' update; very large values for α and β yield very small step sizes and subsequently very slow convergence to the dual maximum, while very small values for α and β yield very large step sizes and subsequently oscillatory and occasionally diverging multipliers' update, requiring a large number of iterations to reach the dual maximum.

However, as widely recognized in relevant studies [93]—[95], the convergence properties of the sub-gradient method are not attractive, as it proceeds to φ^* in a slow and oscillatory fashion. More specifically, the sub-gradient update algorithm zigzags across non-differentiable points of the dual function and generally requires a large number of iterations to reach the dual maximum. Although numerous variation of sub-gradient methods have been developed to tackle its slow and oscillatory convergence, the unfavorable convergence properties have not been sufficiently improved[96].

In contrast to the sub-gradient methods, the *cutting-plane methods*[96-98] utilize the dual function information not only from the latest iteration but also from all previous iterations $\rho = 1, 2, \cdots, r$, therefore exhibiting higher computational requirements per iteration and better convergence properties. More specifically, these methods accumulate the values of the Lagrangian multipliers employed at previous iterations, and the respective sub-gradient vectors and values of the dual function in a "bundle", and use this information to build a piece-wise linear upper approximation $\tilde{\varphi}$ (referred to as *cutting-plane approximation*) of the dual function. The Lagrangian multipliers' values for the next iteration are then determined by maximizing $\tilde{\varphi}$.

The *penalty-bundle* or *proximal-bundle method*[96-98] constitutes another commonly-used multipliers' update approach based on the cutting-plane concept. This method adds a quadratic penalty function in the maximization of $\tilde{\varphi}$, in order to constrain the search space in a region and consequently stabilizes the convergence towards φ^*. This maximisation problem is expressed by (3.78)—(3.79), where γ — the *trust region parameter* of the method — is a positive scalar penalty parameter and $\lambda_c^{[r]}$ — the *centre of gravity* of the method — is a vector of Lagrangian multipliers' values corresponding to one of the previous iterations.

The most popular approach for determining the centre of gravity involves setting it equal to the Lagrangian multipliers' vector corresponding to the iteration having produced the highest value of the actual dual function among the previous iterations[98].

The trust region parameter γ is heuristically determined and its suitable adjustment has a significant impact on the performance of the method[97-98]. A very small value does not introduce a sufficient penalty to the maximization of the approximated dual function. A very large value constrains the search space in a region very close to the previous values of the multipliers, and therefore leads to slow convergence towards the dual maximum.

$$\max_{\lambda^{[r+1]}, \tilde{\varphi}} \tilde{\varphi} - \frac{1}{2}\gamma \|\lambda^{[r+1]} - \lambda_c^{[r]}\|^2 \qquad (3.78)$$

subject to:

$$\tilde{\varphi} \leqslant \varphi(\lambda^{[\rho]}) + (\lambda^{[r+1]} - \lambda^{[\rho]})^T e^{[\rho]}, \quad \forall \rho = 1, 2, \cdots, r \qquad (3.79)$$

The values of the decision variables λ at the solution of the problem (3.78)—(3.79) constitute the trial values of the Lagrangian multipliers for the next iteration. As deduced by (3.79), after every iteration of the method, a new constraint (cut) is added to the above problem, incorporating the dual function information from the latest iteration. Although the cutting-plane method is not an ascending method, as the number of iterations increases, the cutting-plane approximation gets closer to the actual dual function[97], and therefore the solution of (3.78)—(3.79) gets closer to the actual dual function maximum.

Regarding the termination criteria, in references [39]—[40], where the case studies are specifically designed so that the duality gap of the market clearing problem is zero, this iterative algorithm terminates when the norm of the demand-supply imbalances $\left|\sum_{j \in J} d_j^s + D - \sum_{i \in I} g_i^s\right|$ is lower than a pre-specified tolerance. However, in order to accommodate more general cases where the duality gap is not necessarily zero (as a result of participants' non-convexities) and the optimal dual solution does not necessarily satisfy the demand-supply balance constraints[99], the algorithm terminates when the absolute difference in the dual function value between two consecutive iterations is lower than a pre-specified tolerance ε, as expressed by (3.80):

$$|\varphi^{[r+1]} - \varphi^{[r]}| \leqslant \varepsilon \qquad (3.80)$$

The overall structure of the convex hull price computation algorithm employing the Lagrangian multiplier updating methods described above is illustrated in Figure 3.4.

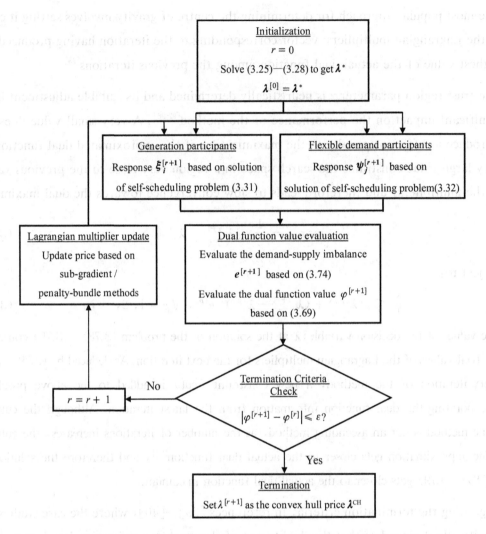

Figure 3.4 Convex hull price computation algorithm flowchart

3.7 Case Studies

3.7.1 Test Data and Implementation

The examined case study involves the demonstration of schedule inconsistency and surplus sub-optimality effects associated with FD non-convexities, as well as the application of both generalized uplift and convex hull pricing approaches to a market with both generation and FD participants, day-ahead horizon and hourly resolution. This study was implemented in FICO™ Xpress[127] on a computer with a 6-core, 3.47 GHz

Intel(R) Xeon(R) X5690 processor and 192GB of RAM.

The market includes 7 generation participants, with fixed costs fc_i^G, linear l_i^G and quadratic q_i^G parameters of their variable cost functions, start-up C_i^u and shut-down C_i^d costs, minimum stable g_i^{min} and maximum g_i^{max} generation limits, ramp-up RU_i and ramp-down RD_i rates, minimum-up UT_i and minimum-down DT_i times, and initial commitment status $u_{i,0}$ and output $g_{i,0}$ given in Table 3.2. It should be noted that the increase in the value of variable cost parameters (l_i^G and q_i^G), the decrease in the non-variable (non-convex) cost parameters (fc_i^G, C_i^u and C_i^d), the decrease in the maximum generation limit (g_i^{max}) and minimum stable generation limit (g_i^{min}), the increase in the ramping rates (RU_i and RD_i), and the decrease in the minimum up/down times (UT_i and DT_i) as we move from the base (cheapest) generator 1 to the peak (most expensive) generator 7, express the range of the technical and economic characteristics of the examined generation merit order.

Table 3.2 Generation participants' characteristics

Generator i	1	2	3	4	5	6	7
$fc_i^G/(£/h)$	18,431	17,005	13,755	9,930	9,900	8,570	7,530
$l_i^G/(£/MWh)$	5.5	30	35	60	80	95	100
$q_i^G/(£/MW^2 h)$	0.0002	0.0007	0.0010	0.0064	0.0070	0.0082	0.0098
$C_i^u/£$	4,000,000	325,000	142,500	72,000	55,000	31,000	11,200
$C_i^d/£$	800,000	28,500	18,500	14,400	12,000	10,000	8,400
g_i^{min}/MW	3,292	2,880	1,512	667	650	288	275
g_i^{max}/MW	6,584	5,760	3,781	3,335	3,252	2,880	2,748
$RU_i/(MW/h)$	1,317	1,152	1,512	1,334	1,951	1,728	2,198
$RD_i/(MW/h)$	1,317	1,152	1,512	1,334	1,951	1,728	2,198
UT_i/h	24	20	16	10	8	5	4
DT_i/h	24	20	16	10	8	5	4
$u_{i,0}$	1	1	1	1	1	0	0
$g_{i,0}/MW$	5,268	4,608	3,025	2,668	2,602	0	0

Furthermore, the market includes 4 CCFD participants and 4 FCFD participants able to forgo and redistribute their demand activities, with parameters given in Table 3.3 and Table 3.4 respectively. Half of the FD participants of each type can be scheduled during night/morning hours, representing domestic FD, and the other half during midday hours, representing commercial/industrial FD. The minimum and maximum power limits of each CCFD are assumed identical at every hour of their scheduling period (and are thus denoted by $d_j^{\min}$ and $d_j^{\max}$ respectively) and zero at the rest of the hours.

Table 3.3 CCFD participants' characteristics

CCFD j	1	2	3	4
B_j^0 / £ mil	0.188	0.418	0.197	0.144
E_j /MWh	2,589	2,784	1,315	958
$d_j^{\min}$/MW	1,262	1,253	394	144
$d_j^{\max}$/MW	1,942	2,088	986	719
Scheduling period	19–8	18–7	9–17	11–18

Table 3.4 FCFD participants' characteristics

FCFD j		5	6	7	8
B_j^0 / £ mil		0.518	0.418	0.134	0.400
T_j^{dur}/h		4	3	1	2
$d_{j,\tau}^{\mathrm{cyc}}$/ MW	$\tau=1$	1,285	581	1,183	1,006
	$\tau=2$	692	628	0	1,660
	$\tau=3$	844	1,575	0	0
	$\tau=4$	630	0	0	0
t_j^{start}		18	19	9	9
t_j^{end}		7	9	17	16

3.7.2 Impacts of Flexible Demand Non-Convexities

Under centralized market clearing, domestic and commercial/industrial FDs fill the inflexible demand's night and midday valleys respectively, flattening significantly the total demand profile, as shown in Figure 3.5. However, these valleys and subsequently the centralized prices (Figure 3.7) at these periods are not completely flattened, due to the minimum power levels of CCFDs and the discrete power levels of FCFDs (subsection 3.4.3).

As depicted in Figures 3.6—3.11, generation and FD participants' self-schedules given the centralized market clearing prices are not consistent with the centralized market clearing schedules. These inconsistencies are associated with generation non-convexities explored in references [33]—[45] and also with both demand flexibility potential. Let us focus on the inconsistencies of the demand side. Considering the ability to redistribute their activities across time, under self-scheduling, FDs concentrate at the lowest-priced

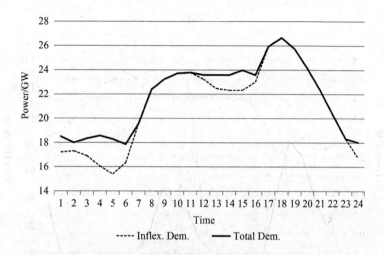

Figure 3.5 Inflexible demand and total demand under centralized market clearing

hours within their scheduling period, which is not consistent with the centralized schedule involving an as-flat-as-possible total demand profile (Figure 3.7). This effect is demonstrated in Figure 3.8 and Figure 3.9 for CCFD 2 and FCFD 6 respectively. Under self-scheduling, the former chooses to acquire its total energy requirements at the two lowest-priced hours within its scheduling period (hour 2 and hour 6), while centralized market clearing schedules it at hour 4 and hour 6. Furthermore, under self-scheduling, FCFD 6 carries out its cycle at hours 4-6, since this leads to lower total payments than carrying it out at hours 3-5 according to centralized market clearing.

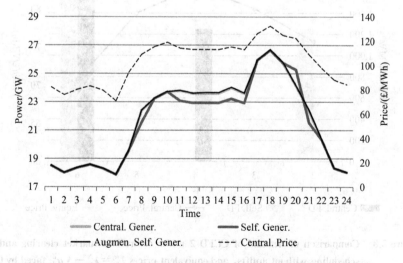

Figure 3.6 Comparison of total generation under centralized market clearing, self-scheduling and augmented self-scheduling

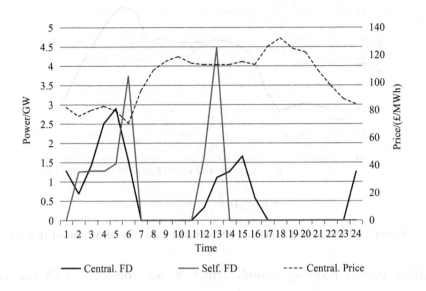

Figure 3.7 Comparison of total FD under centralized market clearing and self-scheduling without uplifts

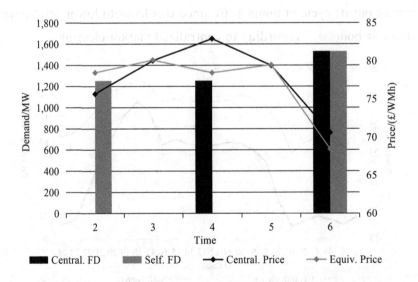

Figure 3.8 Comparison of demand of CCFD 2 under centralized market clearing and self-scheduling without uplifts, and equivalent prices $\lambda_{2t}^{eq} = \lambda_t^N - \Delta \alpha_{2t}^d$ faced by CCFD 2 given the calculated generalized uplift parameters

Chapter 3 Factoring Flexible Demand Non-Convexities in Electricity Markets

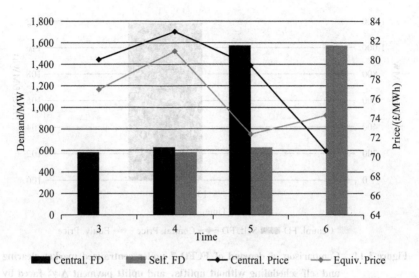

Figure 3.9 Comparison of demand of FCFD 6 under centralized market clearing and self-scheduling without uplifts, and equivalent prices $\lambda_{6t}^{eq} = \lambda_t^N - \Delta\alpha_{6t}^d$ faced by FCFD 6 given the calculated generalized uplift parameters

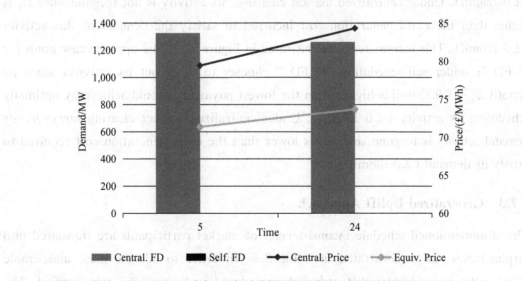

Figure 3.10 Comparison of demand of CCFD 1 under centralized market clearing and self-scheduling without uplifts, and equivalent prices $\lambda_{1t}^{eq} = \lambda_t^N - \Delta\alpha_{1t}^d$ faced by CCFD 1 given the calculated generalized uplift parameters

67

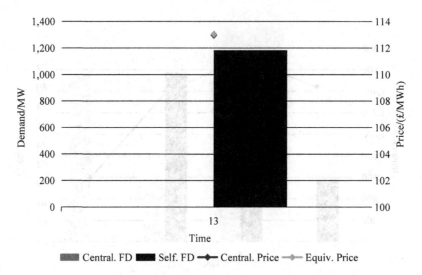

Figure 3.11 Comparison of demand of FCFD 7 under centralized market clearing and self-scheduling without uplifts, and uplift payment $\Delta \gamma_7^d$ faced by FCFD 7 given the calculated generalized uplift parameters

Considering the ability to forgo their demand activities, CCFD 1 chooses to do so under self-scheduling, since its benefit $B_1^0 = £\,0.188\text{mil}$ is lower than the lowest payment it could achieve by optimally scheduling its activity within its feasible scheduling period ($£\,0.189\text{mil}$). Under centralized market clearing, its activity is not forgone since B_1^0 is higher than the extra generation cost incurred to satisfy the demand of this activity ($£\,0.186\text{mil}$). This inconsistency is illustrated in Figure 3.10. The opposite case holds for FCFD 7; under self-scheduling, FCFD 7 chooses to carry out its activity, since its benefit $B_7^0 = £\,0.134\text{mil}$ is higher than the lowest payment it could achieve by optimally scheduling its activity ($£\,0.133\text{mil}$). Under centralized market clearing however, its demand activity is forgone since B_7^0 is lower than the extra generation cost required to satisfy its demand ($£\,0.138\text{mil}$).

3.7.3 Generalized Uplift Approach

The aforementioned schedule inconsistencies of market participants are translated into surplus losses, as demonstrated in Table 3.5. In order to address these undesirable effects, the generalized uplift approach introduced in Section 3.5 was applied. The parameter R associated with the equitable distribution of the total surplus loss compensation was calculated as $R = 99.91\%$ according to (3.56). The new electricity prices were then set according to (3.57) as $\lambda_t^N = \lambda_t^* + 0.095\,£/\text{MWh}$, $\forall\,t$. The generalized uplift parameters' iterative computation algorithm converged after 63 iterations and

Chapter 3 Factoring Flexible Demand Non-Convexities in Electricity Markets

203 s of computational time. As shown in Table 3.5, the total compensation contribution by all participants exactly cancels out the total surplus loss, and the sum of all uplifts is zero.

All FD participants apart from FCFD 7 require some non-zero uplift parameters $\Delta \alpha_{j,t}^d$ and do not require an uplift parameter $\Delta \gamma_j^d$ to address their surplus sub-optimality, as their demand activity is not forgone under centralized market clearing. FCFD 7 requires only a non-zero uplift parameter $\Delta \gamma_j^d$ since its demand activity is forgone under centralized market clearing. The elimination of FD surplus sub-optimality by the calculated uplift parameters is demonstrated by observing the new equivalent electricity prices $\lambda_{jt}^{eq} = \lambda_t^N - \Delta\alpha_{jt}^d$ faced by FD participants in Figures 3.8—3.11.

For CCFD 2 (Figure 3.8), the uplift parameters make the equivalent prices at hours 2 and 4 equal to incentivize CCFD 2 to follow the market clearing solution and self-schedule at hours 4 and 6, as now self-scheduling at hours 2 and 6 (according to the original self-schedule of Figure 3.8) does not bring additional surplus. For FCFD 6 (Figure 3.9), the uplift parameters make λ_{6t}^{eq} lower than λ_t^* at hours 3—5 and higher at hour 6 to incentivize FCFD 6 to follow the market clearing solution and self-schedule its cycle at hours 3—5, as now self-scheduling at hours 4—6 (according to the original self-schedule of Figure 3.9) does not bring additional surplus. For CCFD 1 (Figure 3.10), the uplift parameters reduce significantly λ_{1t}^{eq} with respect to λ_t^* at hours 5 and 24 to incentivize CCFD 1 to follow the market clearing solution and carry out its activity at these two hours, as now forgoing its activity (according to the original self-schedule of Figure 3.10) does not bring additional surplus. Finally, for FCFD 7 (Figure 3.11), the uplift $\Delta \gamma_7^d = £51$ (reward for forgoing its activity) incentivizes FCFD 7 to follow the centralized market clearing solution and forgo its activity, as now carrying out its activity(according to the original self-schedule of Figure 3.11) does not bring additional surplus.

3.7.4 Convex Hull Pricing Approach

The convex hull pricing approach of Section 3.6 was also applied with termination tolerance $\varepsilon = £1$ in (3.80). Both the penalty-bundle algorithm with the trust region parameter set as $\gamma = 800$ in (3.78); and sub-gradient algorithm with $\alpha = 0.1$ and $\beta = 0.05$ in step updating rule (3.77), are employed to solve the dual optimization problem (3.68) and thus determine the convex hull prices. The values of the parameters employed in the penalty-bundle and sub-gradient algorithms are selected on the basis of the guidelines as mentioned in subsection 3.6.3.

Figure 3.12 illustrates the value of the dual function at each of the first 50 iterations of the multipliers' update process for these two algorithms. The sub-gradient method proceeds to the dual maximum in an oscillatory fashion, but demands lowest computational time per iteration; it approaches the dual maximum after approximately 35 iterations and 70 s of computational time. The penalty-bundle method exhibits better convergence properties, but is more computational intensive as it necessitates the solution of a quadratic program at each iteration, it converges to the dual maximum after approximately 20 iterations and computational time of 82 s.

In the context of the convex hull pricing problem examined in this monograph, the number of multipliers' update iterations has a distinctive significance, as it determines the number of message exchanges between the market operator and the decentralized generation and FD participants, and consequently has a major impact on the communication costs associated with a practical implementation of the convex hull pricing mechanism. In this respect, the penalty-bundle method exhibits the most favourable performance comparatively to the sub-gradient method (as it takes a lower of number iterations to reach the dual maximum). Driven by this analysis, the penalty-bundle method is generally deemed as the most favourable multipliers' update method for a larger-scale system.

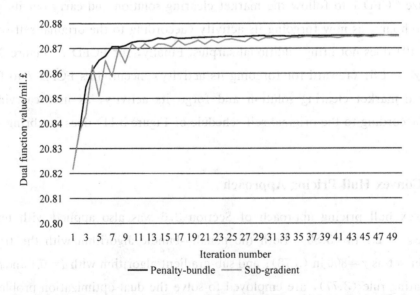

Figure 3.12 Convergence performance of penalty-bundle and sub-gradient methods

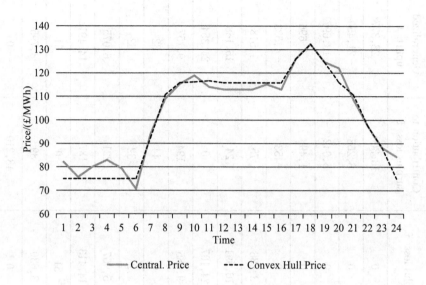

Figure 3.13 Centralized market clearing price and convex hull price

Very interestingly, in contrast to the centralized market clearing prices, convex hull prices are flattened at the night and midday valleys where the FDs are scheduled by centralized market clearing and self-scheduling (Figure 3.13). This flattening effect eliminates surplus sub-optimality associated with the FD ability to redistribute activities across time, as now self-scheduling at different valley hours to the ones determined by centralized market clearing does not bring additional surplus.

As shown in Table 3.6, the positive uplift received by each generation and FD participant exactly cancels out its respective surplus loss, and the total compensation contribution is charged entirely to the inflexible demand, through a negative uplift balancing the positive uplifts received by the rest of the participants. It is worth noting that total surplus loss under convex hull prices (£23,357) is (minimum and) significantly lower than the respective loss under centralized market clearing prices (£77,632). Furthermore, it should also be noted that the primal optimum for the centralized market clearing is £20,898,679 while the dual maximum is found to be £20,875,322 (as illustrated in Figure 3.12), and their difference (namely the duality gap) is £23,357 which coincides with the aforementioned surplus loss under the convex hull prices (subsection 3.6.2).

Table 3.5 Results of generalized uplift approach

Market Participants		Centralized surplus/£	Self-scheduling surplus/£	Augmented self-scheduling surplus/£	Surplus loss/£	Contribution to surplus loss/£	Generalized uplift/£
Generator $i \in I$	1	14,788,400	14,788,400	14,775,100	0	13,300	−28,377
	2	9,188,328	9,188,328	9,180,060	0	8,268	−21,450
	3	5,604,222	5,604,222	5,599,180	0	5,042	−13,697
	4	1,662,733	1,662,733	1,661,240	0	1,493	−7,696
	5	502,784	502,784	502,331	0	453	−4,533
	6	79,703	83,089	83,014	3,386	75	1,518
	7	9,686	26,677	26,653	16,991	24	16,168
CCFD $j \in J^c$	1	−24,107	0	0	24,107	0	24,354
	2	205,627	214,712	214,518	9,085	194	9,157
	3	48,707	48,707	48,663	0	44	82
	4	35,490	35,492	35,460	2	32	62
FCFD $j \in J^f$	5	239,898	245,865	245,644	5,967	221	6,075
	6	193,956	208,369	208,182	14,413	187	14,491
	7	0	51	51	51	0	51
	8	95,153	98,783	98,694	3,630	89	3,795
Inflexible demand		53,577,467	53,529,257	53,577,467	0	48,210	0
Total		53,529,257	53,529,257		77,632	77,632	0

Table 3.6 Results of convex hull pricing approach

Market Participants		Centralized surplus/£	Self-scheduling surplus/£	Augmented self-scheduling surplus/£	Surplus loss/£	Contribution to surplus loss/£	Lump-Sum uplift/£
Generator $i \in I$	1	14,642,469	14,642,469	14,642,469	0	0	0
	2	9,060,677	9,060,677	9,060,677	0	0	0
	3	5,520,419	5,520,419	5,520,419	0	0	0
	4	1,636,037	1,645,527	1,645,527	9,490	0	9,490
	5	504,140	507,526	507,526	3,386	0	3,386
	6	85,005	87,805	87,805	2,800	0	2,800
	7	19,184	20,343	20,343	1,159	0	1,159
CCFD $j \in J^c$	1	−6,477	0	0	6,477	0	6,477
	2	208,840	208,863	208,863	23	0	23
	3	44,992	44,992	44,992	0	0	0
	4	32,782	32,785	32,785	3	0	3
FCFD $j \in J^f$	5	258,820	258,820	258,820	0	0	0
	6	208,838	208,850	208,850	12	0	12
	7	0	0	0	0	0	0
	8	91,238	91,245	91,245	7	0	7
Inflexible demand		53,205,642	53,205,642	53,228,999	0	23,357	−23,357
Total					23,357	23,357	0

3.8 Conclusions

This chapter is dedicated to an important issue of pricing in electricity markets with non-convexities. In this context, previous work has identified non-convexities associated with the generation side of electricity markets and proposed different pricing mechanisms to address surplus sub-optimality. This chapter aims at extending this concept to incorporate the demand side.

Detailed operational models pertaining to two different types of FDs are proposed, capturing the largest part of flexible load models in the related literature. Two different FD flexibility potential are modelled, including the ability to completely forgo demand activities and the ability to redistribute the total electrical energy requirements of activities across time. Non-convexities associated the operation of FD are identified for the first time, including options to forgo demand activities as well as discrete and minimum power levels. The relation of these non-convexities with schedules' inconsistency and surplus sub-optimality effects are demonstrated through simple one- and two-time period examples and a larger case study with day-ahead horizon and hourly resolution.

Both generalized uplift and convex hull pricing approaches are extended to account for FD non-convexities, coordinating surplus-driven market participants to the optimal operational schedule determined by the centralized market clearing problem. Concerning the former, generalized uplift functions for FD participants are proposed, which include FD-specific terms and constitute additional benefits or payments for the FDs. The structure of the generalized uplift function for FDs is not the same with the respective function for generators, due to their distinct non-convexities and resulting surplus sub-optimality effects. The parameters of these functions along with the electricity prices are adjusted to achieve consistency for every FD participant. A new rule is introduced for equitable distribution of the total generators' profit loss and FDs' utility loss compensation among the market participants. Regarding the latter, convex hull prices are calculated as the Lagrangian multipliers optimizing the dual problem of the market clearing problem considering FD participation. It is demonstrated that convex hull prices are flattened at periods when FD is scheduled to eliminate surplus sub-optimality associated with the FD ability to redistribute energy requirements across time.

From a practical implementation point of view, both approaches require solving the market clearing problem (based on the information regarding participants' techno-economic operational characteristics) to determine the centralized optimal schedule, prices, and participants' surpluses. Both approaches then employ a coordinated communication process between the centralized market operator and decentralized participants. More specifically, the former posts incentive signals comprised of prices and uplift payments, the latter then responds to such signals and submits their self-schedules to the former. On the basis of such responses, the former then solves either a minimum discrimination problem (under the generalized uplift approach), or a Lagrangian dual maximization problem (under the convex hull pricing approach) to update the price and uplift signals and post back to the latter. This communication process continues until the final derived prices and uplift payments support the competitive equilibrium solutions. In other words, participants' optimal surplus-maximizing schedules coincide the predetermined centralized optimal schedules (and thus eliminate any surplus sub-optimality effects in the market).

Finally, the introduction of such pricing and compensation mechanisms is likely to spawn gaming strategies by the independent market players. The potential risks associated with gaming under these mechanisms, as well as remedies for mitigating excessive gaming therefore are among the subjects of future research of this chapter.

Nomenclature

Indices and Sets

$t \in T$	Index and set of time periods.
$i \in I$	Index and set of generation participants.
$j \in J$	Index and set of FD participants.
$J^c \subseteq J$	Subset of continuously-controllable FD participants.
$J^f \subseteq J$	Subset of fixed-cycle FD participants.
τ	Index of steps of the cycle of FD participant $j \in J^f$, running from 1 to T_j^{dur}.
r	Index of iterations.
G_i	Operating constraints set of generation participant i.
D_j	Operating constraints set of FD participant j.
$\Xi_i^{[r]}$	Profit cutting plane set of generation participant i at iteration r.
$\Psi_j^{[r]}$	Utility cutting plane set of FD participant j at iteration r.

Variables

λ Vector of electricity prices λ_t (£/MWh).

g_i Vector of power outputs $g_{i,t}$ of generation participant i (MW).

u_i Vector of unit commitment statuses $u_{i,t}$ of generation participant i.

$SU_{i,t}$ Start-up cost of generation participant i at time period t.

$SD_{i,t}$ Shut-down cost of generation participant i at time period t.

d_j Vector of power demands $d_{j,t}$ of FD participant j (MW).

v_j Binary variable expressing whether the activity of FD participant j is forgone ($v_j = 0$ if it is forgone, $v_j = 1$ otherwise).

w_j Vector of binary variables $w_{j,t}$ expressing whether FD participant $j \in J^c$ is active at time period t ($w_{j,t} = 1$ if it is active, $w_{j,t} = 0$ otherwise).

z_j Vector of binary variables $z_{j,t}$ expressing whether the activity of FD participant $j \in J^f$ is initiated at time period t ($z_{j,t} = 1$ if it is initiated, $z_{j,t} = 0$ otherwise).

$\Delta \alpha_i^g$ Vector of uplift parameters $\Delta \alpha_{i,t}^g$ associated with the power output of generation participant i at time period t (£/MWh).

Δc_i^{on} Vector of uplift parameters $\Delta c_{i,t}^{on}$ associated with the "on" commitment status of generation participant i at time period t (£/h).

Δc_i^{off} Vector of uplift parameters $\Delta c_{i,t}^{off}$ associated with the "off" commitment status of generation participant i at time period t (£/h).

$\Delta \alpha_j^d$ Vector of uplift parameters $\Delta \alpha_{j,t}^d$ associated with the power demand of FD participant j at time period t (£/MWh).

$\Delta \gamma_j^d$ Uplift parameter associated with forgoing the activity of FD participant j (£).

Constants

N_T Number of time periods of the horizon.

l_i^G Linear cost coefficient of generation participant i (£/MWh).

q_i^G Quadratic cost coefficient of generation participant i (£/MW²h).

fc_i^G Fixed cost of generation participant i (£/h).

g_i^{min} Minimum power output limit of generation participant i (MW).

g_i^{max} Maximum power output limit of generation participant i (MW).

C_i^u Start-up cost of generation participant i (£).

C_i^d Shut-down cost of generation participant i (£).

R_i^u Ramp-up cost of generation participant i (MW/h).

R_i^d	Ramp-down cost of generation participant i (MW/h).
UT_i	Minimum-up time of generation participant i (h).
DT_i	Minimum-down time of generation participant i (h).
H_i^u	Number of periods generation participant i must be initially online due to its minimum up time constraint.
H_i^d	Number of periods generation participant i must be initially offline due to its minimum down time constraint.
U_i^0	Number of periods generation participant i has been online prior to the first period of the time span (end of period 0).
D_i^0	Number of periods generation participant i has been offline prior to the first period of the time span (end of period 0).
$g_{i,0}$	Initial output of generation participant i (MW).
$u_{i,0}$	Initial unit commitment status of generation participant i.
$\mathbf{D}$	Vector of total inflexible demands D_t (MW).
B_j^0	Benefit associated with the demand activity of FD participant j (£).
$d_{j,t}^{min}$	Minimum power limit of FD participant $j \in J^c$ at time period t (MW).
$d_{j,t}^{max}$	Maximum power limit of FD participant $j \in J^c$ at time period t (MW).
E_j	Energy requirement of the activity of FD participant $j \in J^c$ (MWh).
t_j^{start}	Earliest initiation period of the activity of FD participant $j \in J^f$.
t_j^{end}	Latest termination period of the activity of FD participant $j \in J^f$.
T_j^{dur}	Duration of cycle of FD participant $j \in J^f$ (h).
$d_{j\tau}^{cyc}$	Power demand of FD participant $j \in J^f$ at step τ of its cycle (MW).
n_j	Maximum number of time periods that FD participant j can be active.
del_j^{min}	Minimum delay between two consecutive activations of FD participant j (h).

Functions

C_i	Cost function of generation participant i (£).
B_j	Benefit function of FD participant j (£).
pro_i	Profit function of generation participant i (£).
uti_j	Utility function of FD participant j (£).
U_i^g	Generalized uplift function of generation participant i (£).
U_j^d	Generalized uplift function of FD participant j (£).

Chapter 4
Investigating the Impact of Flexible Demand and Energy Storage on the Exercise of Market Power by Strategic Producers in Imperfect Electricity Markets

4.1 Introduction

Before the 1990s, regulated electricity markets featured vertically-integrated utilities with a monopoly in supply, power providers mainly aimed to minimize the expected cost while maintaining security of supply and were allowed a mandated level of profit[100-102]. However, over the last three decades, the electricity markets have been gradually moved to a more deregulated or liberalized structure, which are characterized by promoting competition among market participants. The intention of establishing an open competitive electricity market is to increase efficiency in the supply of electricity, foster innovation and competition in technologies but also to benefit the consumer through lower electricity prices. On the contrary to what expected, in some deregulated markets, prices have risen sharply higher, the exertion of market power was found to be the underlying cause of the price spikes. The California electricity crisis in 2000 stands for the most notable case highlighting the notorious consequence of market manipulation. According to the estimated statistics, an excess of $ 5.55 billion was paid for electricity from 1998 to 2000 in the deregulated market in California[103]. Identifying the cause of market power and limiting its exercise therefore plays a vital role in improving market efficiency.

The deregulated electricity markets are better described as oligopoly (imperfect competition) rather than perfect competition. In this setting, electricity producers owning a large share of the market or strategically located in the transmission network are able to affect and manipulate the electricity prices and increase their profits beyond the competitive equilibrium levels, through strategic offers. This exercise of market power brings market performance far from the perfect competition equilibrium, with increased price levels and loss of social welfare[46], [100-102].

Previous work on imperfect electricity markets has mainly focused on the generation side, by developing methodologies for optimizing bidding strategies of individual electricity producers as well as investigating market equilibria resulting from the interaction of multiple strategic producers (subsection 4.2.1). From the perspective of mitigating strategic gaming behaviour of the generation side, various remedies have been suggested (subsection 4.2.2). Among others, promoting demand response is regarded as a promising way towards more competitive electricity markets. In this context, the role of the demand side has been previously investigated in terms of the effect of its *own-price elasticity* on producers' ability to exercise market power. However, a large number of researchers have stressed that consumers' flexibility regarding electricity use cannot be fully captured through the concept of own-price elasticity. Instead of simply avoiding using their loads at high price levels, consumers are more likely to shift the operation of their loads from periods of higher prices to periods of lower prices. In other words, load reduction during certain periods is accompanied by a *load recovery effect* during preceding or succeeding periods. This shift of energy demand from high- to low-priced periods drives a demand profile flattening effect. A similar effect is driven by energy storage technologies which are charged during periods of lower prices and are discharged during periods of higher prices. Although numerous recent studies have investigated the impacts of demand shifting[15], [20-27], [65-67], [91] and energy storage[28-32], [59-62], [68-71] on various aspects of power system operation and planning, their roles in imperfect electricity markets has not been explored yet.

The objective of this chapter is to fill this knowledge gap by providing both theoretical and quantitative evidence of the beneficial impact of demand shifting in limiting market power by the generation side. Theoretical explanation of this impact is presented through a price-quantity graph (subsection 4.4.2) in a simplified two-period market. Quantitative analysis is facilitated by a *multi-period equilibrium programming* model of the oligopolistic market setting (Section 4.5), accounting for the time-coupling operational constraints of demand shifting and energy storage, as well as network constraints. Case studies with the developed model on a test market with day-ahead horizon and hourly resolution, operating over a 16-node transmission network quantitatively demonstrate the benefits of demand shifting and energy storage in mitigating the exercise of market power by strategic producers for different scenarios regarding: i) the time-shifting flexibility of the demand side, ii) the size of energy storage, iii) the location of demand shifting and energy storage in the network and iv) network congestion, by employing relevant market power indexes from the literature.

These qualitative and quantitative insights are crucial for both owners and potential investors of generators, in terms of devising suitable strategies in electricity markets considering the increasing penetration of flexible demand and energy storage technologies; as well as regulators and policy makers, in terms of formation of adequate regulatory interventions that exploits the beneficial impact of demand shifting and energy storage to foster more competitive and more efficient electricity markets.

This chapter is organized as follows. Section 4.2 presents a comprehensive literature review regarding modelling approaches of imperfect markets with strategic electricity producers and previously suggested market power mitigation approaches. Section 4.3 outlines models of generation, demand and storage market participants. Section 4.4 provides a theoretical explanation of the beneficial impact of demand side and energy storage on market power. Section 4.5 details the formulation of the oligopolistic market model. Case studies and quantitative results are presented in Section 4.6. Finally, Section 4.7 discusses conclusions of this chapter.

4.2 Literature Review

4.2.1 Modelling Imperfect Markets with Strategic Electricity Producers

As discussed in the previous section, electricity market has gradually evolved towards a liberalized structure. Opportunities follow this market restructuring that allow electricity producers to attain more profits with the attempt to manipulate market prices. In this context, previous work has widely employed the *Game-theoretic models* (GTMs)[47-55] for optimizing bidding strategies of individual electricity producers as well as investigating market oligopolistic equilibria resulting from the interaction of multiple strategic producers.

The GTMs, also called equilibrium models, that simulate an electricity market as a game in which each player competes against each other to determine its bidding strategy, and the payoff (i.e. net earnings) attained by each player are dependent to the strategies of other players, and the market outcome is simulated from the perspective of players' *mutual interactions*. The optimal economic equilibria of the market can be determined through *Nash Equilibrium* (NE) *Theory*, where a NE expresses a strategy combination of all players that no player sees any reason to deviate its decision given the rest of the players do not deviate from their decisions. GTM is deemed to most accurately reflect the actual behaviour of players in the real electricity markets and is adopted widely for simulating and analysing market power[104-105].

According to the competition level in the market, GTMs can be classified as three types: *Bertrand*, *Cournot* and *Supply Function Equilibrium* (SFE)[104]. In a Bertrand competition model, the competition level is the highest, strategic producers compete to decide at what price they sell their production. A Cournot competition model features lowest competition level, in which producers compete to decide how much they produce. Finally, in an SFE model, producers compete through simultaneous choice of supply functions rather than compete in prices or quantities alone. The SFE model constitutes a good compromise between the Cournot and Bertrand models with moderate competition level as well as the equilibria price. Additionally, generation competition is more akin to supply function competition as the realistic market models typically accept price-quantity pairwise bids for the sale of electricity by the producers[50], [102], [105]. In this chapter, individual strategic producers will compete in the market to choose the supply function offer that maximizes its profit.

Regarding the solution techniques of the GTM models, references [106]—[107] developed methods to find the explicit *reaction function* of each strategic producer. However, these models lack scalability, as they are difficult to extend to the case with large number of variables, where analytically unravelling the explicit reaction functions may be challenging. Alternative approaches are proposed in references [47]—[55], where the strategic behaviour of each producer is initially modelled as a bi-level optimization problem. The upper level determines the optimal offering strategy maximizing the profit of the producer, and is subject to the lower level problem representing the market clearing process. These two problems are interrelated since the optimal offer strategies determined by the upper-level problem affect the objective function of the lower-level problem, whereas the market clearing outcome determined by the lower-level problem affects the objective function of the upper-level problem. In order to solve this bi-level problem, the latter is converted to a single-level *Mathematical*

Program with Equilibrium Constraints (MPECs)[①], by replacing the lower level problem by its equivalent *Karush-Kuhn-Tucker* (KKT) conditions[108-109]. However, the resultant MPEC problem is highly non-linear and non-convex. For computational tractability, different linearization approaches have been put forward. A binary expansion approach is adopted in reference [51], where the variable of the market price is discretized with the introduction of additional binary variables in order to linearize the objective function of the MPEC problem. In reference [53], the same non-linearity is eliminated by exploiting the *strong duality theorem* and making use of some KKT conditions. In both cases, the non-linear MPEC is subsequently transformed to a *Mixed-Integer Linear Program* (MILP) form, which can be solved efficiently using available commercial solvers.

The MPEC formulation expresses the decision-making process of a single strategic producer. In order to determine the oligopolistic market equilibrium under the participation of multiple electricity producers, recent work has employed two distinct approaches. Under the first one, the KKT conditions pertaining to the MPECs of each producer are derived and the concatenation of all conditions constitutes a single optimization problem, known as *Equilibrium Program with Equilibrium Constraints* (EPEC). Under the second one, known as *diagonalization*, each producer solves iteratively their MPEC problems — given the offering strategies of the rest of the producers as determined in the previous iteration — until the offering strategies of all producers remain constant with respect to the previous iteration (subsection 4.5.4).

To sum up, the bi-level optimization model captures the interrelationship between the profit maximization problem of each strategic producer and the market clearing problem of the market operator. The impact of producers' strategic offers on the formation of market clearing prices and quantities is endogenously modelled in the optimization. Furthermore, due to the specific features of the electricity industry, imperfect electricity markets with strategic producers are better described as oligopoly (subsection 2.5.1). In this regard, the GTM provides a suitable modelling framework for analysing the oligopolistic equilibrium resulting from reconciling the conflicting interests of multiple

---

① MPEC approaches are widely adopted to devise optimal bidding strategies for market participants in imperfect electricity markets. This approach was initially proposed in reference [47] for modelling strategic behaviour of electricity producers. Other studies used MPECs models for different purposes such as optimal investment of strategic producers in reference [111], the strategic bidding of large consumers in references [57]—[58], and the strategic bidding of large energy storage units in reference [62] (as also employed in Chapter 5).

independent producers.

This chapter develops an *equilibrium programming* model of the oligopolistic market setting. A bi-level optimization model is developed for the decision-making process of each strategic producer competing with rival producers in a market with supply function offers (subsection 4.3.1). The upper level represents the profit maximization problem of the producer and is subject to the limits of the strategic offer variables and the lower level problem. The latter represents the market clearing process where the market operator maximizes the social welfare accounting for generation and demand sides' characteristics and electric network constraints. This problem is solved after converting it to an MPEC and linearizing the latter through suitable techniques. The oligopolistic market equilibria resulting from the interaction of multiple independent strategic producers are determined by employing an iterative diagonalization algorithm.

4.2.2 Generation Market Power Mitigation

As discussed in Section 4.1, the abuse of market power of electricity producers results in adverse consequences including loss of social welfare and significant increase in consumer payments. Market power mitigation therefore among the major concerns in electricity market design and operation.

A wide range of literature has identified various remedies to limit market power exercise[100], [102], [110-115]. They are briefly discussed as follows:

- *Limitation of market share*. This method involves encouraging the separation and divestiture of the dominant generation firms to foster market competition. One of the earliest examples of this was in the UK, where the conventional generation units of the formerly state-owned monopoly were split into new companies, and were later mandated to further divest their assets[100].

- *Ease of entry*. Threat of entry is an effective deterrent to the exertion of market power. Encouraging new participants by lowering or removing barriers to entry is also a useful way to animate open competitive markets. The barriers to entry in generation include environment regulations, scarcity of locations, long amortization periods, and construction time, etc.

- *Expansion of transmission system*. Network congestion favours market power exercise by strategically-located producers. Therefore, adequate transmission expansion serves to decrease concentration of generation by expanding the geographic market area over which producers compete.

- *Contract-based methods*. Long-term contracts also play a vital role in market power mitigation. Under such methods, dominant electricity producers are required to sell a certain amount of their production by virtue of long-term contracting at a regulated rate. This scheme provides great certainty and stability of revenues to producers by removing their long-term exposure to electricity price volatility. Intuitively, the more of a producer's output is covered by contracts, the less energy traded in the market, and the less impetus it will have to behave strategically in the market.

- *Price caps and bidding restrictions*. In addition to the structural solutions (aiming at fostering competition through changes of market structure); secondary solutions involve imposing price caps on market clearing prices and control on the bidding of generators. Temporary price caps provide an upper limit to the pool market price, whereas bidding control restricts producers' ability to bid far away from their actual marginal cost.

Aside from the aforementioned remedies, improving demand side responsiveness is also regarded as a promising way towards market power mitigation. Previous work investigating the role of the demand side in this context has demonstrated that *own-price elasticity* of demand dwindles electricity producers' ability to exercise market power[21], [116-120], as demand is reduced at high market prices and thus limits the volume of electricity sold by strategic producers. A theoretical explanation of this effect is presented in references [21], [116]. Authors in references [117]—[118] employ an SFE model to determine the market equilibria with different levels of demand's price elasticity and define a number of market power indexes in order to quantitatively analyse the impact of elasticity. In reference [119], the same authors model the effect of two demand response programs, namely *time-of-use* (TOU) *pricing* and *economic load response program* (ELRP), on the price elasticity of demand and subsequently assess their impacts on the extent of exercised market power. Finally, an agent-based electricity market model is employed in reference [120] to assess the benefits of different elasticity levels in limiting market power. As discussed in Section 4.1 however, the concept of own-price elasticity cannot accurately capture consumers' flexibility, as the latter mainly involves shifting of loads' operation in time. Furthermore, the single-period modelling framework adopted by previous work [117]—[119] cannot quantify the inter-temporal dependencies between the impact of demand shifting/energy storage and the extent of exercised market power by the generation side.

This chapter aims at tackling the above challenges. The market clearing process (represented by the lower level problem) takes into account the time-coupling operational constraints of demand shifting and energy storage. In other words, the impacts of the latter on i) the formation of market clearing outcome, subsequently on ii) the decision-making of each strategic producer (since the lower level problem will send feedback to each producer regarding how its strategy affects the market clearing price and generation dispatch), and eventually on iii) the oligopolistic market equilibrium are all accurately captured in the optimization process. Consequently, the developed multi-period equilibrium programming model constitutes a suitable framework to achieve the objective of this chapter (Section 4.1).

4.3 Modelling Market Participants

As discussed in Section 4.5, the strategic decision-making of each generation participant is modelled through a bi-level optimization problem whose lower level represents market clearing process. In order to solve this bi-level problem, the lower level problem is replaced by its optimality conditions, which is enabled by the *continuity* and *convexity* (Appendix A) of the lower level problem. It is therefore assumed that all market participants are without non-convex operational characteristics. Nevertheless, the market clearing process still takes into account the time-coupling operational characteristics of the demand side and energy storage, but they are modelled without any binary or integer variables (subsection 2.5.2).

4.3.1 Strategic Generation Participants

For presentation clarity reasons, we assume that each generation participant i owns a single generation unit. The non-variable generation costs (e.g. fixed costs) and minimum stable generation limit are omitted in the formulation of generation cost function and constraints, as the combination of the former and the unit commitment statuses creates non-convexities in the model (subsection 3.4.2). A simplified yet representative model is employed, where each of the generation units is characterized by the quadratic variable cost function (4.1), linear marginal cost function (4.2) and the maximum generation capability (4.3) respectively.

$$C_{i,t}(g_{i,t}) = l_i^G g_{i,t} + q_i^G (g_{i,t})^2 \tag{4.1}$$

$$C'_{i,t}(g_{i,t}) = l_i^G + 2 q_i^G g_{i,t} \tag{4.2}$$

$$0 \leqslant g_{i,t} \leqslant g_i^{\max}, \forall t \tag{4.3}$$

The quadratic variable cost function introduces non-linearity into the optimization model of Section 4.5, adding to the complexity of the problem. Therefore, it is often desirable to approximate the quadratic cost characteristics (4.1) by a piecewise linear interpolation, comprising of a number of generation blocks. The piecewise linear function is indistinguishable from the quadratic function if sufficiently large number of segments are used[90]. Since each segment of the cost curve is linear, each segment of the marginal cost curve is constant, which leads to a stepwise (or stairwise[90]) linear marginal cost curve. This model follows the common practice of most market designs where producers/consumers submit their price-quantity offer/bids in the form of stepwise supply/demand curves[109]. The cost, marginal cost function and output limits of each block b are expressed by (4.4), (4.5) and (4.6) respectively:

$$C_{i,t,b}(g_{i,t,b}) = \lambda_{i,b}^G g_{i,t,b} \tag{4.4}$$

$$C'_{i,t,b}(g_{i,t,b}) = \lambda_{i,b}^G \tag{4.5}$$

$$0 \leqslant g_{i,t,b} \leqslant g_{i,b}^{\max}, \forall t \tag{4.6}$$

As shown in Figure 4.1, the supply offer of generation unit i consists of several production blocks, derived from a piecewise linearization of its quadratic cost curve. Thus, each block includes a generation quantity and its corresponding price offer.

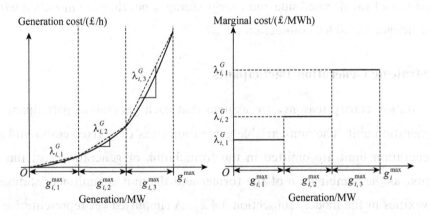

Figure 4.1 Quadratic, piecewise linear cost function, and stepwise marginal cost function

The profit of generation participant i at time period t is expressed as the difference between the revenue it receives from the sale of the energy it produces at the locational marginal price (LMP), and the cost of producing this energy.

$$GP_{i,t}(\lambda_{(n:\,i\in I_n),\,t},\,g_{i,t,b}) = \sum_{b}[(\lambda_{(n:\,i\in I_n),\,t} - \lambda_{i,b}^{G})\,g_{i,t,b}] \quad (4.7)$$

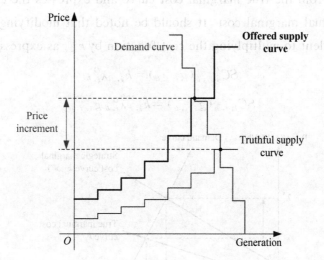

Figure 4.2　Economic withholding and the associated impact on market price

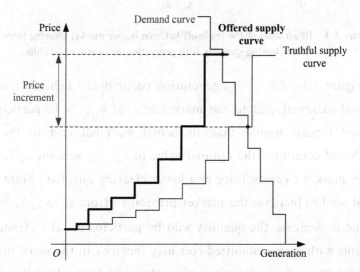

Figure 4.3　Physical withholding and the associated impact on market price

Strategic generation participants can exercise market power through submitting offers higher than their actual marginal costs (i.e. *economic withholding*) and hope this will increase the market price (Figure 4.2); or offering less than their actual generation capacity to the market (i.e. *physical withholding*) and hope the market price rises enough to compensate for the reduced volume sold (Figure 4.3)[46],[89].

Following the model employed in references [48]—[50], [71], [117] the *strategic marginal cost function* is expressed by (4.8), where the value of the decision variable

87

$k_{i,t} \geqslant 1$ represents the strategic behaviour of generation participant i at time period t, it varies the offer from the true marginal cost curve and expresses the extent by which it misreports its actual marginal cost. It should be noted that modifying the offer in this manner is equivalent to multiplying the cost function by $k_{i,t}$ as expressed by (4.9).

$$SC'_{i,t,b}(g_{i,t,b}) = k_{i,t} \lambda^G_{i,b} \qquad (4.8)$$

$$SC_{i,t,b}(g_{i,t,b}) = k_{i,t} \lambda^G_{i,b} g_{i,t,b} \qquad (4.9)$$

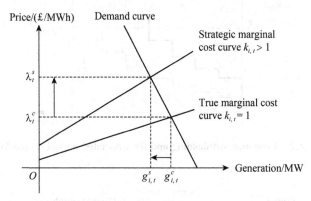

Figure 4.4 Illustration of the trade-off between higher market clearing price and lower clearing quantity with generation economic withholding

As shown in Figure 4.4, if $k_{i,t} = 1$, generation participant i behaves competitively and reveals its actual marginal cost to the market at t. If $k_{i,t} > 1$, participant i behaves strategically and reports higher than its actual marginal cost to the market at t. Participant i should determine the optimal value of $k_{i,t}$ by accounting for the trade-off between higher market clearing price and lower clearing quantity. More specifically, a higher $k_{i,t}$ will tend to increase the market price at t (from λ^c_t to λ^s_t), but at the same time it will tend to decrease the quantity sold by participant i at t (from $g^c_{i,t}$ to $g^s_{i,t}$), since participants with lower submitted cost may replace i in the merit order and/or the demand side and the energy storage may employ self-price elasticity or time-shifting flexibility to reduce net demand at t. Therefore, the participants' net earnings will increase (and thereby exert market power) only if the price it receives per unit sold rises enough to compensate for any decrease in the total quantity it sells.

4.3.2 Demand Participants

The execution of the electrical load operation generally provides some *satisfaction* or *usefulness* to the consumer, following the model employed in references [117]—[119], the benefit obtained by the demand participants at each time period is expressed through a quadratic, non-decreasing and concave function (4.10). The *marginal benefit* or *willingness to pay* is thus expressed through a linear downward-sloping function (4.11) which captures the effect of demand's own-price elasticity. That is, as the demand level increases consumers are willing to pay a lower price, or equivalently, as the market price increases the demand requested by consumers is reduced. The limits in the requested demand level at each time period are expressed by equation (4.12). The maximum price $l^D_{j,t}$, represents the *value of lost load* (VoLL)[120] that measures the average price per megawatt that consumers are willing to pay to prevent being disconnected without notice. The VoLL, the slope of the marginal benefit function and the maximum power limit are time-specific parameters, capturing the differentiated preferences of consumers across different time periods[57].

$$B_{j,t}(d_{j,t}) = l^D_{j,t} d_{j,t} - q^D_{j,t}(d_{j,t})^2 \qquad (4.10)$$

$$B'_{j,t}(d_{j,t}) = l^D_{j,t} - 2 q^D_{j,t} d_{j,t} \qquad (4.11)$$

$$0 \leqslant d_{j,t} \leqslant d^{max}_{j,t}, \quad \forall t \qquad (4.12)$$

In line with subsection 4.3.1, the quadratic benefit function (4.10) is approximated by a piecewise linear benefit function, consisting of a number of demand blocks. The benefit, marginal benefit, and demand limits of each block c are expressed by (4.13), (4.14) and (4.15) respectively:

$$B_{j,t,c}(d_{j,t,c}) = \lambda^D_{j,t,c} d_{j,t,c} \qquad (4.13)$$

$$B'_{j,t,c}(d_{j,t,c}) = \lambda^D_{j,t,c} \qquad (4.14)$$

$$0 \leqslant d_{j,t,c} \leqslant d^{max}_{j,t,c}, \quad \forall t \qquad (4.15)$$

A common feature of the electricity markets is the lack of *price responsiveness* caused by the peculiar characteristics of the electricity, such as non-storability and lack of good substitutes. As reported in reference [121], a statistics of £16,940 £/MWh was established as the overall UK national average VoLL for domestic, small and medium enterprises (SME) customers. This figure suggests that the consumers are not likely to reduce their comfort and convenience in exchange for a cut on their electricity bills by a few percentage. This weak elasticity facilitates the exercise of market power by generation participants. As discussed in subsection 4.3.1, if the demand is very inelastic (i.e. the demand curve exhibits a very steep slope), the price increase driven by market power will be large enough to offset the decrease in quantity sold, the generation participants therefore are able to attain higher profits.

However, a large number of researchers have stressed that consumers' flexibility regarding electricity use cannot be fully captured through the concept of own-price elasticity. Instead of simply avoiding using their loads at high price levels, consumers are more likely to shift the operation of their loads from periods of higher prices to periods of lower prices[18],[21],[73], reflecting the time-coupling operational characteristics of the demand side. This ability of the demand side to redistribute its energy requirements across time will be enhanced with the penetration of various flexible demand technologies, such as smart-charging electric vehicles and smart domestic appliances, and advanced metering, control and communication technologies in consumers' premises, envisaged by the *Smart Grid* paradigm[56].

In this chapter, the aforementioned *time-shifting flexibility potential* of the demand side is expressed by (4.16)—(4.18). The variable $d_{j,t}^{sh}$ represents the change of the demand of demand participant j with respect to the baseline level $\sum_c d_{j,t,c}$ at time period t due to load shifting, taking negative values when demand is moved away from t and positive values when demand is moved towards t. Constraint (4.17) ensures that demand shifting is energy neutral within the examined time horizon i.e. the total size of demand reductions is equal to the total size of demand increases (load recovery), assuming that demand shifting does not involve energy gains or losses. Constraint (4.18) expresses the limits of demand change at each time period due to load shifting as a ratio α_j ($0 \leqslant \alpha_j \leqslant 1$) of the baseline demand; $\alpha_j = 0$ implies that participant j does not exhibit any time-shifting flexibility, while $\alpha_j = 1$ implies that the whole demand of participant j can be shifted in time.

$$d'_{j,t} = \sum_c d_{j,t,c} + d_{j,t}^{sh}, \quad \forall t \tag{4.16}$$

$$\sum_t d_{j,t}^{sh} = 0 \tag{4.17}$$

$$-\alpha_j \times \sum_c d_{j,t,c} \leqslant d_{j,t}^{sh} \leqslant \alpha_j \times \sum_c d_{j,t,c}, \quad \forall t \tag{4.18}$$

The utility of the demand participant j at time period t is given by (4.19). It indicates the surplus of the demand side as a benefit of electricity consumption minus the cost of electricity use. While the energy payment (second term) depends on the final demand after any potential load shifting, the benefit (first term) is assumed to depend only on the baseline demand; this assumption expresses the flexibility of the consumers to shift the operation of some of their loads without compromising the satisfaction they experience.

$$DU_{j,t}(\lambda_{(n;\,j\in J_n),\,t},\,d_{j,t,c},\,d'_{j,t}) = \sum_c \lambda^D_{j,t,c} d_{j,t,c} - \lambda_{(n;\,j\in J_n),\,t} d'_{j,t} \quad (4.19)$$

To sum up, the developed operational model (4.13)—(4.19) captures both the own-price elasticity and time-shifting flexibility of the demand side. A steep downward-sloping marginal benefit function is used to model the low elasticity of the demand side. Inter-temporal operational constraints are employed to represent consumers' flexibility to reschedule their energy requirements across time, and such flexibility potential is realised with no disruptions to the consumers' utilities.

It should be stressed that strategic demand participants can strategically increase their surplus by submitting bids lower than their actual marginal benefit (thereby exercise market power), as demonstrated in references [57]—[58]. However, since the focus of this chapter is on the market power potential of generation participants, demand participants are assumed competitive entities revealing their actual economic and physical characteristics to the market.

4.3.3 Energy Storage

Since the focus of this chapter is to investigate the impact of energy storage on the market power exercised by the generation participants (Section 4.1), a single energy storage unit of significant size is assumed for the sake of the analysis. As such, the storage charging/discharging schedules are able to alter the system net demand to affect the market prices, and thus affect the ability of generation participants to manipulate market prices through strategic offering. As mentioned in Section 2.2, the storage unit is considered as an independent market player, which is neither part of a storage-solar[68] nor storage-wind[69] coalition, nor is part of the portfolio of a generation company[59], nor owned by different market players[60]. While those are viable cases, modelling them is beyond the scope of this chapter.

The inter-temporal operational characteristics of the energy storage unit are expressed by (4.20)—(4.24). Constraint (4.20) expresses that the energy contained in the storage at the end of current time step is calculated as the energy content at the previous time step, plus energy charged during the current time step, minus energy discharged during the current time step considering charging and discharging losses. Constraint (4.21) corresponds to the minimum and maximum limits of the energy content which are related to its maximum depth of discharge and state of charge ratings respectively. Constraints

(4.22)—(4.23) represent the limits of storage's electrical power input that are dependent on the electrical power capacity of the storage battery. For the sake of simplicity, the storage energy content at the start and the end of the market's temporal horizon are assumed equal, ensuring energy conservation.

$$E_t = E_{t-1} + \eta^c s_t^c - s_t^d / \eta^d, \quad \forall t \qquad (4.20)$$

$$E^{\min} \leqslant E_{r,t} \leqslant E^{\max}, \quad \forall t \qquad (4.21)$$

$$0 \leqslant s_t^c \leqslant s^{\max}, \quad \forall t \qquad (4.22)$$

$$0 \leqslant s_t^d \leqslant s^{\max}, \quad \forall t \qquad (4.23)$$

$$E_0 = E_{N_T} \qquad (4.24)$$

As will be thoroughly analysed in Chapter 5, large energy storage units can exercise market power by withholding their power capacity. However, since the focus of this chapter is on the market power potential of generation participants, energy storage is assumed to behave competitively, revealing its true operational characteristics to the market.

4.4 Theoretical Analysis of Impact of Demand Side and Energy Storage on Market Power

In this section, we explore the impact of the demand side on the strategic behaviours of the generation participants with theoretical analysis. In the first subsection, the impact of demand own-price elasticity on the strategic offering decision is assessed in a single-period market example. Next, on the basis of the time-coupling characteristics presented in the previous section, the impact of the flexibility enabled by demand shifting and energy storage on the extent of exercised market power is illustrated in a simplified two-period market example.

4.4.1 Impact of Demand Own-Price Elasticity

As discussed in subsection 4.3.2, the very low elasticity of the demand facilitates the exercise of market power by generation participants as the price increment driven by market power will be large enough to offset the quantity decrement. To this end, authors in references [117]—[119] argue that the increased own-price elasticity of demand deters the market power exertion. Figure 4.5 illustrates this mitigation effect in a single-period market example.

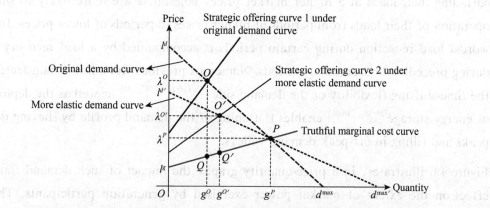

Figure 4.5 Illustration of impacts of the increased demand own-price elasticity on the strategic bidding behaviour of the producers

As illustrated in Figure 4.5, the intersection between the truthful marginal cost curve and the original demand curve (dashed blue) identifies the market equilibrium point P under perfect competition. The increased demand own-price elasticity can be achieved by pivoting the original demand curve counter-clockwise around point P (Figure 4.5). Under the more elastic demand curve (dashed red), the consumer will not buy as much as he/she did with the original demand curve at the same market price. The consumer will be willing to buy the same quantity, but at a lower price instead. In the other words, consumer becomes more responsive to price variations.

It should be noted that the optimal offering curve of the strategic producer is different under demand elasticity, and the reason behind this is explained as follows. Under the more elastic demand curve, demand is reduced at higher market price, this in turn limits the quantity sold by the strategic producer. As a result, it tends to be more profitable for the strategic producer to bid a lower mark-up in order to sell more quantity to the market. This is represented in Figure 4.5 by the adjustment of strategy from offering curve 1 to offering curve 2, which is closer to the actual marginal cost curve, and the production is higher at the oligopoly equilibrium (i.e. $g^{O'} > g^{O}$). As a consequence of the changes in both demand and supply curves, the market clearing price at the oligopoly equilibrium decreases from λ^{O} to $\lambda^{O'}$, and the producer's profit (denoted by the area of a trapezoid) is reduced from the area $\lambda^{O} l^{g} QO$ to the area $\lambda^{O'} l^{g} Q'O'$.

4.4.2 Impacts of Demand Shifting and Energy Storage

As discussed in subsection 4.3.2, consumers' flexibility regarding their electricity use cannot be captured through the concept of demand own-price elasticity. Instead of curtailing their loads at a higher market price, consumers are more likely to shift the operation of their loads from periods of higher prices to periods of lower prices. In other words, load reduction during certain periods is accompanied by a load recovery effect during preceding or succeeding periods. Numerous previous work has demonstrated that the time-shifting flexibility of the demand side[15], [20-27], [65], [91] as well as the deployment of energy storage[28-32], [70-71] enables flattening of the demand profile by shaving demand peaks and filling in off-peak demand valleys.

Figure 4.6 illustrates, in a price-quantity graph, the impact of such demand flattening effect on the extent of market power exercised by generation participants. The two curves represent in a simplified fashion the aggregated competitive and strategic offer curves of the generation side. While most of the capacity is offered from base generation within a relatively narrow price band, the price for the intermediate and peaking generation increases sharply[21], [46]. Furthermore, the price intercept and the slope of the strategic curve are higher than the respective parameters of the competitive curve (subsection 4.3.1). As mentioned in subsection 4.3.2, electrical demand exhibits very high VoLL and very low elasticity, it is therefore assumed that the demand side is perfectly inelastic (i.e. the down-sloping demand curve is simplified to a vertical demand line).

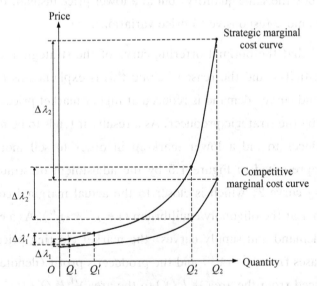

Figure 4.6 Impacts of demand shifting and energy storage on generation participants' ability to manipulate market prices

As discussed in subsection 4.3.1, the strategic action of the generation participants through economic withholding has a positive impact on the generation profits, since it results in increase of price levels. Conversely, it also has a negative impact, since it reduces the quantity sold by the generation participants. However, the assumed weak elasticity of the demand ensures that the resultant price increase will be large enough to offset the reduction in the volume sold. In other words, the extent of market power (as indicated by the increase of generation profit) is dictated by the increment of price levels. Consequently, the proceeding analysis focuses on the impact of demand shifting and energy storage on the market price increment, and the respective impact on the increase of generation profit can be equivalently implied.

The intersections of the marginal cost curves with the vertical demand lines determine the market clearing prices in the respective cases. The price increment $\Delta\lambda$ represent the increase of the market clearing prices driven by the exercise of market power in the respective cases. As demonstrated in Figure 4.6, this price increase is much higher during the peak period due to the increasing slope of the offer curve.

Demand shifting and energy storage reduce the peak demand from Q_2 to Q_2' and increase the off-peak demand from Q_1 to Q_1'. As a result, this reduces the price increment at the peak period from $\Delta\lambda_2$ to $\Delta\lambda_2'$ while increases it at the off-peak period from $\Delta\lambda_1$ to $\Delta\lambda_1'$ (Figure 4.6). Although the peak demand reduction is equal to (in the case of demand shifting, due to the assumed energy neutrality constraint (4.17)) or even lower than (in the case of energy storage, due to the charging and discharging losses included in constraint (4.20)) the off-peak demand increase, i.e. $Q_2 - Q_2' \leqslant Q_1' - Q_1$, the resultant price increment reduction at the peak period is higher than its increase at the off-peak period, i.e. $\Delta\lambda_2 - \Delta\lambda_2' > \Delta\lambda_1' - \Delta\lambda_1$, due to the larger slope of the strategic marginal cost curve. Overall, demand shifting and energy storage limit the ability of strategic generation participants to manipulate market prices (for their own benefit).

This effect also applies to the resulting increase of generation profit (as quantitatively explored in subsection 4.6.2), and implies that flexibility enabled by demand shifting and energy storage results in an overall reduction of the extent of market power exercised by the generation side.

4.5 Modelling Oligopolistic Electricity Markets with Demand Shifting and Energy Storage

As discussed in subsection 2.5.1, the optimization models used in the literature to represent perfect and imperfect market are fundamentally different. In order to represent perfect markets, it suffices to adopt a single optimization problem representing the market clearing problem (solved by the market operator) that determines the dispatch of each participant that maximises the social welfare (given the truthful bids of the participants).

In order to represent imperfect market however, a more complex bi-level optimization model, comprising of two coupled optimization problems, is used in the literature. The upper level problem represents the decision-making problem of individual market participant and determines the strategy of the participant reported to the market (depending on the participant under study, this strategy could associated with its economic and technical characteristics) that maximizes their individual surplus. The lower level problem still represents the market clearing problem solved by the market operator; however, the difference with respect to perfect markets is the market operator generally perceives untruthful technical and/or economic characteristics from the participants, thus the optimal market outcome will be inherently dependent on these misreported characteristics. These two optimization problems are interrelated since the strategy determined by the upper level problem affects the clearing dispatch of the participant and clearing prices determined by the lower level problem; and the latter is feedback to the upper level problem and affect individual participant's surplus. This bi-level optimization model is adopted in Chapter 4 and Chapter 5 for the decision-making of strategic generation and energy storage participants.

4.5.1 Bi-Level Optimization Model

Following the approach employed in references [47]—[55], the decision-making of each strategic generation participant is modelled through a bi-level optimization model. As illustrated in Figure 4.7, the bi-level model is constitutive of an upper level (UL) problem as well as a lower level (LL) problem. In the UL, each participant behaves strategically through the offering decisions maximizing its generation profit. The UL problem is subject to the LL problem that represents endogenously the market clearing process, maximizing the *perceived* (since the generation participants do not generally report their actual marginal costs (subsection 4.3.1)) social welfare, accounting for the time-coupling operational characteristics of the demand shifting and energy storage (subsection 4.3.2 and subsection 4.3.3), and employing a DCOPF model to account for the effect of the network.

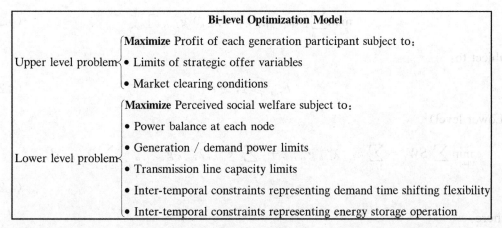

Figure 4.7 Bi-level structure of the decision-making problem of a strategic generation participant

Each generation participant optimizes its offering strategy in the UL while receiving feedback (i.e. LMPs and generation dispatch) from the LL regarding how its strategy affects the market outcome. Therefore, the offering decisions are variables in the UL problem while considered as parameters in the LL problem.

The UL and LL problems are coupled as illustrated in Figure 4.8. On the one hand, the offering strategy determined by the upper level problem affects the objective function of the LL problem, and thus affect the market outcomes. On the other hand, the market clearing price and generation dispatches determined by the lower level problem influence the objective function (i.e. the profit of the generation participant) of the upper level problem.

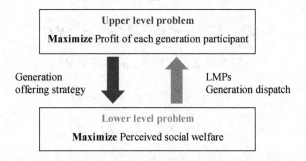

Figure 4.8 Interrelation between the upper level and lower level problems

Considering the participation of demand shifting and energy storage in the market, the formulation of the proposed bi-level model of a generation participant i competing with rival participants with supply function offers is given by equation (4.25) — equation (4.40).

(Upper level)

$$\max_{k_{i,t}} \sum_{t,b} \left[(\lambda_{(n;\,i\in I_n),\,t} - \lambda^G_{i,b}) g_{i,t,b} \right] \quad (4.25)$$

subject to:

$$k_{i,t} \geqslant 1, \; \forall t \quad (4.26)$$

(Lower level)

$$\min_{V^{\text{LL-P}}} \sum_t SW_t = \sum_{t,b} k_{i,t} \lambda^G_{i,b} g_{i,t,b} + \sum_{i^-,t,b} k_{i^-,t} \lambda^G_{i^-,b} g_{i^-,t,b} - \sum_{j,t,c} \lambda^D_{j,t,c} d_{j,t,c}$$

$$(4.27)$$

where:

$$V^{\text{LL-P}} = \{ g_{i,t,b},\, d_{j,t,c},\, d^{\text{sh}}_{j,t},\, s^c_t,\, s^d_t,\, E_t,\, \theta_{n,t} \} \quad (4.28)$$

subject to:

$$\sum_{(j\in J_n),c} d_{j,t,c} + \sum_{j\in J_n} d^{\text{sh}}_{j,t} + (s^c_t - s^d_t)_{n_s} - \sum_{(i\in I_n),b} g_{i,t,b} + \sum_{m\in M_n} \frac{(\theta_{n,t} - \theta_{m,t})}{x_{n,m}} = 0 : \lambda_{n,t},\; \forall n,\; \forall t$$

$$(4.29)$$

$$0 \leqslant g_{i,t,b} \leqslant g^{\max}_{i,b} : \mu^-_{i,t,b},\, \mu^+_{i,t,b},\; \forall i,\; \forall t,\; \forall b \quad (4.30)$$

$$0 \leqslant d_{j,t,c} \leqslant d^{\max}_{j,t,c} : \nu^-_{j,t,c},\, \nu^+_{j,t,c},\; \forall j,\; \forall t,\; \forall c \quad (4.31)$$

$$\sum_t d^{\text{sh}}_{j,t} = 0 : \xi_j,\; \forall j \quad (4.32)$$

$$-\alpha_j \sum_c d_{j,t,c} \leqslant d^{\text{sh}}_{j,t} \leqslant \alpha_j \sum_c d_{j,t,c} : \pi^-_{j,t},\, \pi^+_{j,t},\; \forall j,\; \forall t \quad (4.33)$$

$$E_t = E_{t-1} + \eta^c s^c_t - s^d_t / \eta^d : \rho_t,\; \forall t \quad (4.34)$$

$$E^{\min} \leqslant E_t \leqslant E^{\max} : \sigma^-_t,\, \sigma^+_t,\; \forall t \quad (4.35)$$

$$0 \leqslant s^c_t \leqslant s^{\max} : \varphi^-_t,\, \varphi^+_t,\; \forall t \quad (4.36)$$

$$0 \leqslant s^d_t \leqslant s^{\max} : \chi^-_t,\, \chi^+_t,\; \forall t \quad (4.37)$$

$$E_0 = E_{N_T} : \psi \quad (4.38)$$

$$-F^{\max}_{n,m} \leqslant \frac{\theta_{n,t} - \theta_{m,t}}{x_{n,m}} \leqslant F^{\max}_{n,m} : \beta^-_{n,m,t},\, \beta^+_{n,m,t},\; \forall n,\; \forall m \in M_n,\; \forall t \quad (4.39)$$

$$-\pi \leqslant \theta_{n,t} \leqslant \pi : \gamma^-_{n,t},\, \gamma^+_{n,t},\; \forall n,\; \forall t \quad (4.40)$$

$$\theta_{1,t} = 0 : \delta_t,\; \forall t \quad (4.41)$$

The objective function (4.25) of the UL problem constitutes the profit of the generation participant i. This problem is subject to the UL constraints pertaining to the limits of the strategic offer variables (4.26) (subsection 4.3.1) and the LL problem (4.27)—(4.40) representing the market clearing process at each time period, maximizing the perceived social welfare (4.27). It should be noted that the strategic offer of the participant i is a UL decision variable but treated as given values within the LL problem. This suggests that once all the participants submit their strategic offers, $k_{i,t}$, $\forall t$ of participant i and $k_{i-,t}$, $\forall i-$, $\forall t$ of its competitors, the market operator clears the market through (4.27)—(4.40) considering all the collected offers. The LL problem is subject to nodal demand-supply balance constraints (4.29) (the Lagrangian multipliers of which constitute the locational marginal prices (LMPs) at each node and each period), generation and demand power limits (4.30)—(4.31), operational constraints capturing the time-coupling characteristics of the demand side (4.32)—(4.33) and energy storage (4.34)—(4.38). Limits on transmission line capacities and voltage angles of nodes are enforced through constraints (4.39)—(4.40), respectively. Finally, constraint (4.41) identifies $n=1$ as the reference node.

4.5.2 MPEC Formulation

In order to solve the bi-level model (4.25)—(4.41) of each strategic generation participant, it is convenient to convert it into a single-level Mathematical Program with Equilibrium Constraints (MPEC) optimization problem. The transformation is carried out by replacing the lower level problem with its optimality conditions as illustrated in Figure 4.9. This transformation is enabled by the assumed continuity and convexity of the lower level problem[108]—[109].

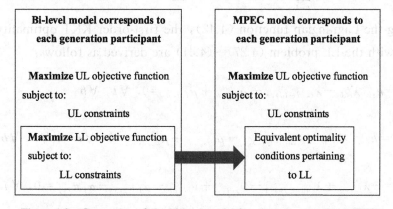

Figure 4.9 Conversion of the bi-level model into its corresponding MPEC

According to references [55], [108]—[109], the equivalent optimality conditions pertaining to the LL problem can be formulated through two alternative approaches: *Karush-Kuhn-Tucker* (KKT) conditions and *primal-dual transformation*.

KKT conditions

Concerning the first approach, the derivation of the KKT conditions associated with the LL problem (4.27)—(4.41), starts with expressing the corresponding Lagrangian function L:

$$L = \sum_{t,b} k_{i,t} \lambda^G_{i,b} g_{i,t,b} + \sum_{i-,t,b} k_{i-,t} \lambda^G_{i-,b} g_{i-,t,b} - \sum_{j,t,c} \lambda^D_{j,t,c} d_{j,t,c} +$$

$$\sum_{n,t} \lambda_{n,t} \left[\sum_{(j \in J_n),c} d_{j,t,c} + \sum_{j \in J_n} d^{sh}_{j,t} + s^c_t - s^d_{t\,ns} - \sum_{(i \in I_n),b} g_{i,t,b} + \sum_{m \in M_n} \frac{(\theta_{n,t} - \theta_{m,t})}{x_{n,m}} \right] -$$

$$\sum_{i,t,b} [\mu^-_{i,t,b} g_{i,t,b} - \mu^+_{i,t,b}(g_{i,t,b} - g^{max}_{i,b})] -$$

$$\sum_{j,t,c} [\nu^-_{j,t,c} d_{j,t,c} - \nu^+_{j,t,c}(d_{j,t,c} - d^{max}_{j,t,c})] +$$

$$\sum_{j,t} \xi_j d^{sh}_{j,t} - \sum_{j,t} \pi^-_{j,t}(\alpha_j \sum_c d_{j,t,c} + d^{sh}_{j,t}) + \sum_{j,t} \pi^+_{j,t}(d^{sh}_{j,t} - \alpha_j \sum_c d_{j,t,c}) +$$

$$\sum_t \rho_t (E_t - E_{t-1} - \eta^c s^c_t + s^d_t / \eta^d) +$$

$$\sum_t [\sigma^-_t (E^{min} - E_t) + \sigma^+_t (E_t - E^{max})] + \psi(E_0 - E_{N_T}) -$$

$$\sum_t [\varphi^-_t s^c_t - \varphi^+_t (s^c_t - s^{max})] - \sum_t [\chi^-_t s^d_t - \chi^+_t (s^d_t - s^{max})] -$$

$$\sum_{n,(m \in M_n),t} \beta^-_{n,m,t} \left(\frac{\theta_{n,t} - \theta_{m,t}}{x_{n,m}} + F^{max}_{n,m} \right) + \sum_{n,(m \in M_n),t} \beta^+_{n,m,t} \left(\frac{\theta_{n,t} - \theta_{m,t}}{x_{n,m}} - F^{max}_{n,m} \right) -$$

$$\sum_{n,t} \gamma^-_{n,t}(\theta_{n,t} + \pi) + \sum_{n,t} \gamma^+_{n,t}(\theta_{n,t} - \pi) + \sum_t \delta_t \theta_{1,t} \qquad (4.42)$$

Considering the Lagrangian function (4.42), the first-order KKT optimality conditions associated with the LL problem (4.27)—(4.41) are derived as follows:

$$\frac{\partial L}{\partial g_{i,t,b}} = k_{i,t} \lambda^G_{i,b} - \lambda_{(n:\,i \in I_n),t} - \mu^-_{i,t,b} + \mu^+_{i,t,b} = 0, \ \forall t, \ \forall b \qquad (4.43)$$

$$\frac{\partial L}{\partial g_{i-,t,b}} = k_{i-,t} \lambda^G_{i-,b} - \lambda_{(n:\,i- \in I_n),t} - \mu^-_{i-,t,b} + \mu^+_{i-,t,b} = 0, \ \forall i-, \ \forall t, \ \forall b \qquad (4.44)$$

$$\frac{\partial L}{\partial d_{j,t,c}} = -\lambda^D_{j,t,c} + \lambda_{(n:\,j \in J_n),t} - \nu^-_{j,t,c} + \nu^+_{j,t,c} - \alpha_j \pi^-_{j,t} - \alpha_j \pi^+_{j,t} = 0, \ \forall j, \ \forall t, \ \forall c$$

$$(4.45)$$

$$\frac{\partial L}{\partial d_{j,t}^{sh}} = \lambda_{(n,\,j\in J_n),\,t} + \xi_j - \pi_{j,t}^- + \pi_{j,t}^+ = 0, \quad \forall j, \forall t \qquad (4.46)$$

$$\frac{\partial L}{\partial s_t^c} = \lambda_{n_s,t} - \eta^c \rho_t - \varphi_t^- + \varphi_t^+ = 0, \quad \forall t \qquad (4.47)$$

$$\frac{\partial L}{\partial s_t^d} = -\lambda_{n_s,t} + \rho_t/\eta^d - \chi_t^- + \chi_t^+ = 0, \quad \forall t \qquad (4.48)$$

$$\frac{\partial L}{\partial E_t} = \rho_t - \rho_{t+1} - \sigma_t^- + \sigma_t^+ = 0, \quad \forall t < N_T \qquad (4.49)$$

$$\frac{\partial L}{\partial E_{N_T}} = \rho_{N_T} - \sigma_{N_T}^- + \sigma_{N_T}^+ - \psi = 0 \qquad (4.50)$$

$$\frac{\partial L}{\partial \theta_{n,t}} = \sum_{m\in M_n} \frac{\lambda_{n,t} - \lambda_{m,t}}{x_{n,m}} + \sum_{m\in M_n} \frac{\beta_{n,m,t}^+ - \beta_{m,n,t}^+}{x_{n,m}} -$$
$$\sum_{m\in M_n} \frac{\beta_{n,m,t}^- - \beta_{m,n,t}^-}{x_{n,m}} + \gamma_{n,t}^+ - \gamma_{n,t}^- + (\delta_t)_{n=1} \qquad (4.51)$$
$$= 0, \quad \forall n, \forall t$$

$$\sum_{(j\in J_n),c} d_{j,t,c} + \sum_{j\in J_n} d_{j,t}^{sh} + (s_t^c - s_t^d)_{n_s} - \sum_{(i\in I_n),b} g_{i,t,b} + \sum_{m\in M_n} \frac{(\theta_{n,t} - \theta_{m,t})}{x_{n,m}} = 0, \quad \forall n, \forall t \qquad (4.52)$$

$$\sum_t d_{j,t}^{sh} = 0, \quad \forall j \qquad (4.53)$$

$$E_t = E_{t-1} + \eta^c s_t^c - s_t^d/\eta^d, \quad \forall t \qquad (4.54)$$

$$E_0 = E_{N_T} \qquad (4.55)$$

$$\theta_{1,t} = 0, \quad \forall t \qquad (4.56)$$

$$0 \leq \mu_{i,t,b}^- \perp g_{i,t,b} \geq 0, \quad \forall i, \forall t, \forall b \qquad (4.57)$$

$$0 \leq \mu_{i,t,b}^+ \perp (g_{i,b}^{max} - g_{i,t,b}) \geq 0, \quad \forall i, \forall t, \forall b \qquad (4.58)$$

$$0 \leq \nu_{j,t,c}^- \perp d_{j,t,c} \geq 0, \quad \forall j, \forall t, \forall c \qquad (4.59)$$

$$0 \leq \nu_{j,t,c}^+ \perp (d_{j,t,c}^{max} - d_{j,t,c}) \geq 0, \quad \forall j, \forall t, \forall c \qquad (4.60)$$

$$0 \leq \pi_{j,t}^- \perp (d_{j,t}^{sh} + \alpha_j \sum_c d_{j,t,c}) \geq 0, \quad \forall j, \forall t \qquad (4.61)$$

$$0 \leq \pi_{j,t}^+ \perp (\alpha_j \sum_c d_{j,t,c} - d_{j,t}^{sh}) \geq 0, \quad \forall j, \forall t \qquad (4.62)$$

$$0 \leq \sigma_t^- \perp (E_t - E^{min}) \geq 0, \quad \forall t \qquad (4.63)$$

$$0 \leq \sigma_t^+ \perp (E^{max} - E_t) \geq 0, \quad \forall t \qquad (4.64)$$

$$0 \leqslant \varphi_t^- \perp s_t^c \geqslant 0, \ \forall t \tag{4.65}$$

$$0 \leqslant \varphi_t^+ \perp (s_r^{\max} - s_t^c) \geqslant 0, \ \forall t \tag{4.66}$$

$$0 \leqslant \chi_t^- \perp s_t^d \geqslant 0, \ \forall t \tag{4.67}$$

$$0 \leqslant \chi_t^+ \perp (s_r^{\max} - s_t^d) \geqslant 0, \ \forall t \tag{4.68}$$

$$0 \leqslant \beta_{n,m,t}^- \perp \left(F_{n,m}^{\max} + \frac{\theta_{n,t} - \theta_{m,t}}{x_{n,m}} \right) \geqslant 0, \ \forall n, \ \forall m \in M_n, \ \forall t \tag{4.69}$$

$$0 \leqslant \beta_{n,m,t}^+ \perp \left(F_{n,m}^{\max} - \frac{\theta_{n,t} - \theta_{m,t}}{x_{n,m}} \right) \geqslant 0, \ \forall n, \ \forall m \in M_n, \ \forall t \tag{4.70}$$

$$0 \leqslant \gamma_{n,t}^- \perp (\pi + \theta_{n,t}) \geqslant 0, \ \forall n, \ \forall t \tag{4.71}$$

$$0 \leqslant \gamma_{n,t}^+ \perp (\pi - \theta_{n,t}) \geqslant 0, \ \forall n, \ \forall t \tag{4.72}$$

The structure of the KKT conditions (4.43)—(4.72) is outlined below:

Equality constraints (4.43)—(4.51) are the *stationary conditions* derived from differentiating the Lagrangian function L with respect to the primal variables in set $V^{\text{LL-P}}$ identified in (4.28). Equality constraints (4.52)—(4.56) are the primal equality constraints (4.29), (4.32), (4.34), (4.38) and (4.41) in the LL problem. Complementary slackness conditions (4.57)—(4.72) are associated with the inequality constraints (4.30)—(4.31), (4.33), (4.35)—(4.37) and (4.39)—(4.40) in the LL problem. Note that the obtained KKT conditions (4.43)—(4.72) are in the first place *first-order necessary optimality conditions*, which are also *sufficient conditions of optimality* since the considered LL problem is linear and thus convex.

It should be noted that the LL problem (as well as similar problems in the existing literature [47]—[55], [117]—[118]) neglects the complex unit commitment constraints of the generation side, due to their inherent inability to deal with binary decision variables. As mentioned in subsection 6.6.2, future work aims at exploring mathematical techniques enabling (approximate) incorporation of these complex constraints in the developed model without deteriorating significantly its computational performance.

Primal-dual transformation

Regarding the second approach, the first step is to derive the corresponding dual problem of the LL problem (4.27)—(4.41). The dual problem is given by (4.73)—(4.83) below:

$$\max_{V^{\text{LL-D}}} -\sum_{t,b} \mu^+_{i,t,b} g^{\max}_{i,b} - \sum_{i-,t,b} \mu^+_{i-,t,b} g^{\max}_{i-,b} - \sum_{j,t,c} \nu^+_{j,t,c} d^{\max}_{j,t,c} + (\psi - \rho_1) E_0 +$$

$$\sum_t (\sigma^-_t E^{\min} - \sigma^+_t E^{\max} - \varphi^+_t s^{\max} - \chi^+_t s^{\max}) - \sum_{n,(m \in M_n),t} (\beta^-_{n,m,t} + \beta^+_{n,m,t}) F^{\max}_{n,m} -$$

$$\sum_{n,t} (\gamma^-_{n,t} + \gamma^+_{n,t}) \pi \qquad (4.73)$$

where:

$$V^{\text{LL-D}} = \{\lambda_{n,t}, \mu^-_{i,t,b}, \mu^+_{i,t,b}, \nu^-_{j,t,c}, \nu^+_{j,t,c}, \xi_j, \pi^-_{j,t}, \pi^+_{j,t}, \rho_{k,t}, \sigma^-_{k,t},$$
$$\sigma^+_{k,t}, \varphi^-_{k,t}, \varphi^+_{k,t}, \chi^-_{k,t}, \chi^+_{k,t}, \psi_k, \beta^-_{n,m,t}, \beta^+_{n,m,t}, \gamma^-_{n,t}, \gamma^+_{n,t}, \delta_t\} \qquad (4.74)$$

subject to:

$$(4.43)-(4.51) \qquad (4.75)$$

$$\mu^-_{i,t,b}, \mu^+_{i,t,b} \geq 0, \quad \forall i, \forall t, \forall b \qquad (4.76)$$

$$\nu^-_{j,t,c}, \nu^+_{j,t,c} \geq 0, \quad \forall j, \forall t, \forall c \qquad (4.77)$$

$$\pi^-_{j,t}, \pi^+_{j,t} \geq 0, \quad \forall j, \forall t \qquad (4.78)$$

$$\sigma^-_t, \sigma^+_t \geq 0, \quad \forall t \qquad (4.79)$$

$$\varphi^-_t, \varphi^+_t \geq 0, \quad \forall t \qquad (4.80)$$

$$\chi^-_t, \chi^+_t \geq 0, \quad \forall t \qquad (4.81)$$

$$\beta^-_{n,m,t}, \beta^+_{n,m,t} \geq 0, \quad \forall n, \forall m \in M_n, \forall t \qquad (4.82)$$

$$\gamma^-_{n,t}, \gamma^+_{n,t} \geq 0, \quad \forall n, \forall t \qquad (4.83)$$

Considering the primal problem (i.e. LL problem) (4.27)—(4.41), and the corresponding dual problem (4.73)—(4.83), the required optimality conditions pertaining to the market clearing process are expressed as given by (4.84)—(4.86). Constraint (4.84) represents the strong duality equality that expresses the equivalence between the values of the primal objective function (4.27) and dual objective function (4.73) at the optimal solution. It should be noted that the optimality conditions (4.84)—(4.86) resulting from prima-dual transformation are equivalent to the KKT conditions (4.43)—(4.72) derived earlier (refer to reference [108] for detailed mathematical proof).

$$\sum_{t,b} k_{i,t} \lambda^G_{i,b} g_{i,t,b} + \sum_{i-,t,b} k_{i-,t} \lambda^G_{i-,b} g_{i-,t,b} - \sum_{j,t,c} \lambda^D_{j,t,c} d_{j,t,c}$$

$$= -\sum_{t,b} \mu^+_{i,t,b} g^{\max}_{i,b} - \sum_{i-,t,b} \mu^+_{i-,t,b} g^{\max}_{i-,b} - \sum_{j,t,c} \nu^+_{j,t,c} d^{\max}_{j,t,c} + (\psi - \rho_1) E_0 +$$

$$\sum_t (\sigma^-_t E^{\min} - \sigma^+_t E^{\max} - \varphi^+_t s^{\max} - \chi^+_t s^{\max}) - \sum_{n,(m \in M_n),t} (\beta^-_{n,m,t} + \beta^+_{n,m,t}) F^{\max}_{n,m} -$$

$$\sum_{n,t} (\gamma^-_{n,t} + \gamma^+_{n,t}) \pi \qquad (4.84)$$

$$(4.29)-(4.41) \tag{4.85}$$

$$(4.75)-(4.83) \tag{4.86}$$

Comparing the KKT and primal-dual approaches, the first approach includes a set of complementarity slackness conditions as a part of the optimality conditions. As explained in subsection 4.5.3, such conditions can be linearized with *Big-M method* by introducing an additional set of auxiliary binary variables into the model. The second approach is more succinct by including the strong duality equality as a part of the optimality conditions. However, the latter adds non-linearity in the form of the product between two continuous variables. Therefore, considering the two observations above, the first approach is adopted in this chapter to represent the optimality conditions pertaining to the LL problem. However, it necessitates the derivation of the strong duality equality (through primal-dual transformation) in order to linearize the final MPEC optimization model (subsection 4.5.3).

Following the above discussion, the resultant single level MPEC of generation participant i is given by (4.87)—(4.90) below.

$$\max_{V^{MPEC}} \sum_{t,b} \left[(\lambda_{(n; i \in I_n), t} - \lambda_{i,b}^G) g_{i,t,b} \right] \tag{4.87}$$

where:

$$V^{MPEC} = \{k_{i,t}, V^{LL-P}, V^{LL-D}\} \tag{4.88}$$

subject to:

$$k_{i,t} \geqslant 1, \forall t \tag{4.89}$$

$$(4.43)-(4.72) \tag{4.90}$$

The objective function of the MPEC is identical to the objective function of the UL problem. The set of decision variables (4.88) of the resultant MPEC model includes the decision variables of the UL and the LL problem as well as the Lagrangian multipliers associated with the constraints of the LL problem.

4.5.3 MILP Formulation

It is evident that the initial MPEC model (4.87)—(4.90) is non-linear and thus any solution obtained by commercial solvers is not guaranteed to be globally optimal. The objective of this subsection is to transform this MPEC to a mixed-integer linear problem (MILP) which can be efficiently solved using currently available branch-and-cut solvers.

Two types of non-linearities are presented in the above MPEC model. The first one involves the bilinear terms $\sum_{t,b} \lambda_{(n:\,i\in I_n),\,t}\, g_{i,t,b}$ in the objective function (4.87), which involves a product of the market clearing price and generation schedule variables. The linearization approach proposed in reference [53] is employed in order to replace these terms with a linear expression. This approach exploits the strong duality equality (4.84) and some of the KKT conditions (4.43)—(4.72). By multiplying both sides of (4.43) by $g_{i,t,b}$, summing for every t and b and rearranging some terms we get:

$$\sum_{t,b} k_{i,t} \lambda_{i,b}^G g_{i,t,b} = \sum_{t,b} (\lambda_{(n:\,i\in I_n)}\, g_{i,t,b} + \mu_{i,t,b}^-\, g_{i,t,b} - \mu_{i,t,b}^+\, g_{i,t,b}) \quad (4.91)$$

By making use of (4.57), equation (4.91) becomes:

$$\sum_{t,b} k_{i,t} \lambda_{i,b}^G g_{i,t,b} = \sum_{t,b} (\lambda_{(n:\,i\in I_n)}\, g_{i,t,b} - \mu_{i,t,b}^+\, g_{i,t,b}) \quad (4.92)$$

By making use of (4.58), equation (4.92) becomes:

$$\sum_{t,b} k_{i,t} \lambda_{i,b}^G g_{i,t,b} = \sum_{t,b} (\lambda_{(n:\,i\in I_n)}\, g_{i,t,b} - \mu_{i,t,b}^+\, g_{i,b}^{\max}) \quad (4.93)$$

By substituting (4.93) into (4.84) and rearranging some terms we get:

$$\sum_{t,b} \lambda_{(n:\,i\in I_n)}\, g_{i,t,b} = \sum_{j,t,c} (\lambda_{j,t,c}^D\, d_{j,t,c} - \nu_{j,t,c}^+\, d_{j,t,c}^{\max}) -$$
$$\sum_{i-,t,b} (k_{i-,t}\, \lambda_{i-,b}^G\, g_{i-,t,b} + \mu_{i-,t,b}^+\, g_{i-,b}^{\max}) + (\psi - \rho_1) E_0 +$$
$$\sum_{t} (\sigma_t^-\, E^{\min} - \sigma_t^+\, E^{\max} - \varphi_t^+\, s^{\max} - \chi_t^+\, s^{\max}) -$$
$$\sum_{n,(m\in M_n),t} (\beta_{n,m,t}^- + \beta_{n,m,t}^+)\, F_{n,m}^{\max} - \sum_{n,t} (\gamma_{n,t}^- + \gamma_{n,t}^+)\pi$$

$$(4.94)$$

Therefore, the bilinear terms $\sum_{t,b} \lambda_{(n:\,i\in I_n),\,t}\, g_{i,t,b}$ in the objective function (4.87) of the MPEC problem can be replaced with the expression in the right side of (4.94), which is linear. The resulting objective function of the MILP formulation is:

$$\max_V \sum_{t,c} (\lambda_{t,c}^D\, d_{t,c} - \nu_{t,c}^+\, d_{t,c}^{\max}) - \sum_{i-,t,b} (k_{i-,t}\, \lambda_{i-,b}^G\, g_{i-,t,b} + \mu_{i-,t,b}^+\, g_{i-,b}^{\max}) + (\psi - \rho_1) E_0 +$$
$$\sum_{t} (\sigma_t^-\, E^{\min} - \sigma_t^+\, E^{\max} - \varphi_t^+\, s^{\max} - \chi_t^+\, s^{\max}) - \sum_{n,(m\in M_n),t} (\beta_{n,m,t}^- + \beta_{n,m,t}^+)\, F_{n,m}^{\max} - \sum_{n,t} (\gamma_{n,t}^- +$$
$$\gamma_{n,t}^+)\pi - \sum_{t,b} \lambda_{i,b}^G\, g_{i,t,b} \quad (4.95)$$

The second non-linearity involves the bilinear terms in the complementarity conditions (4.57)—(4.72). The linearization approach proposed in reference [122] is employed to replace these conditions with a set of mixed-integer linear conditions. If we denote each of the complementarity conditions as $0 \leqslant \mu \perp p \geqslant 0$, then this approach replaces it with the conditions $\mu \geqslant 0$, $p \geqslant 0$, $\mu \leqslant \omega M^{\mu}$, and $p \leqslant (1-\omega) M^{p}$, where ω is an auxiliary binary variable, M^{μ} and M^{p} are large enough positive constants. The set of decision variables of the MILP formulation includes the set (4.88) as well as the auxiliary binary variables introduced for linearizing (4.57)—(4.72). For example, constraint (4.57) can be replaced by its mixed-integer equivalence (4.96)—(4.100) as below.

$$\bar{\mu}_{i,t,b} \geqslant 0, \quad \forall i, \forall t, \forall b \tag{4.96}$$

$$g_{i,t,b} \geqslant 0, \quad \forall i, \forall t, \forall b \tag{4.97}$$

$$\bar{\mu}_{i,t,b} \leqslant \omega_{i,t,b} M^{\mu}, \quad \forall i, \forall t, \forall b \tag{4.98}$$

$$g_{i,t,b} \leqslant (1-\omega_{i,t,b}) M^{p}, \quad \forall i, \forall t, \forall b \tag{4.99}$$

$$\omega_{i,t,b} \in \{0,1\}, \quad \forall i, \forall t, \forall b \tag{4.100}$$

It should be stressed that the effectiveness and computational time of the proposed MILP model is significantly dependent on the values of M^{μ} and M^{p}. Thus the selection of the values of M^{μ} and M^{p} is crucial and must be adequate. Several guidelines on how to select a suitable value are outlined in references [53], [55], [122]. Specifically, M^{μ} and M^{p} should be large enough so as not to impose additional incorrect bounds on the variables (causing infeasibility). However, a too large value tends to impede the convergence of the branch-and-cut solvers[55]. Consequently, naively selecting big value for M^{μ} and M^{p} is certainly not an efficient solution, ideally we desire to determine the minimum value for these constants. On the basis of the physical limits of the primal variables, the minimum value of M^{p} can be easily calculated. For instance, considering the above constraints (4.96)—(4.100), if $\omega_{i,t,b}=1$, then the variable $g_{i,t,b}$ is equal to zero as determined by constraints (4.97) and (4.99); while if $\omega_{i,t,b}=0$, then $g_{i,t,b} \leqslant M^{p}$, simultaneously considering the physical limits (4.30) of the generation variable, this implies that the minimum value of M^{p} is $g_{i,b}^{\max}$.

However, the selection of the values for the constants associated with the dual variables is challenging since the dual variables do not exhibit an unambiguous physical bound, thus it presents a great challenge to mathematically determine the minimum required values for M^{μ} (as did for M^{p}). To address this issue, a trial-and-error approach is generally adopted to tune the value so that the constraint $\mu \leqslant \omega M^{\mu}$ is inactive if $\omega = 1$, and at the same time M^{μ} is small enough for the sake of algorithm convergence.

For computational tractability, a heuristic approach has been implemented and proven to be beneficial for the selection of M^μ value. This approach first involves the solution of the market clearing problem (4.27)—(4.41) with $k_{i,t}=1$, $\forall i$, $\forall t$ and then store the optimal dual variables μ^* associated with relevant inequality constraints (4.30)—(4.31), (4.33), (4.35)—(4.37) and (4.39)—(4.40). Subsequently, the value of the each relevant constant is calculated as $M^\mu = (\mu^* + 1) \times \gamma$, where 1 is used as an offset in case $\mu^* = 0$, and γ constitutes the parameter of this approach. Then, we set $\gamma = 100$, and then solve the model. Next, we check the numerical results to validate if the linearized (4.95) representation of generation profit matches the non-linear one (4.87) and if each complementarity condition holds. If so, the value of γ is reduced, and then the model is resolved. This tuning-up process continues until infeasibility occurs. Although this method is somewhat heuristic, it is more efficient than the naïve approach as it makes use of the information of the actual dual variable values associated with the inequality constraints in the market clearing process under competitive generation behaviours.

4.5.4 Determining the Oligopolistic Market Equilibrium

The above MPEC/MILP formulation expresses the decision-making process of a single generation participant. In order to determine the oligopolistic market equilibrium reconciling conflicting interests of multiple independent generation participants, recent work has employed two distinct approaches.

Under the first one, the equilibrium in the market is determined by concatenating all the MPECs from all strategic generation participants, which constitutes an EPEC. Each MPEC is replaced by its equivalent KKT conditions[55] to characterize strong stationarity solutions of the EPEC, as illustrated in Figure 4.10. The main drawback of this approach is its modelling and computational complexity, mainly associated with the significant non-convexities and non-linearities (such as *over-lapping complementarity conditions*) of the EPEC formulation[50], [52], [54-55]. In this context, relevant work has employed different linearization techniques which however sacrifice the accuracy and optimality of the obtained solution.

Under the second one, known as diagonalization, each of the participants solves iteratively their MPEC problem — given the offering strategies of the rest of the participants as determined in the previous iteration — until the offering strategies of all participants remain constant (within some tolerance) with respect to the previous iteration. This state corresponds to a market equilibrium since none of the participants

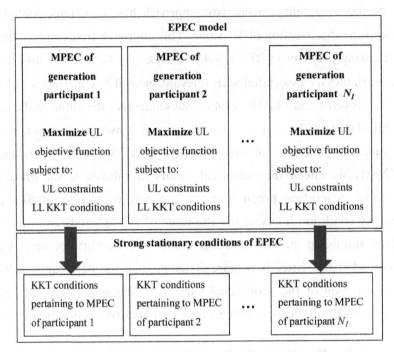

Figure 4.10 EPEC model and KKT conditions pertaining to EPEC

can increase their profit by unilaterally modifying their offering strategies. In terms of computational tractability, the second approach has a clear advantage over the EPEC approach due to the fact that each non-linear MPEC (corresponds to each strategic producer's decision-making process) is fully transferred to its equivalent MILP formulation, which means that any solution obtained by available commercial solvers is guaranteed to be global optimal. Therefore, this iterative diagonalization approach is employed in this chapter for the determination of the NE.

This approach is presented in Figure 4.11. At each iteration, each of the generation participants determines their individual offering decisions by solving the MPEC formulated in subsection 4.5.2 and subsection 4.5.3, given the decisions of the rest of the generation participants as determined in the previous iteration (In the first iteration, each of the participants will assume its rivals submit offers equal to their actual marginal costs). This iterative procedure terminates when the decisions of all players remain constant (within some tolerance ε) with respect to the previous iteration — implying that a NE has been reached — or a maximum number of iterations has been executed.

As discussed in the literature, existence and uniqueness of market equilibria are not generally guaranteed[47],[105]. In order to deal with this challenge, the above

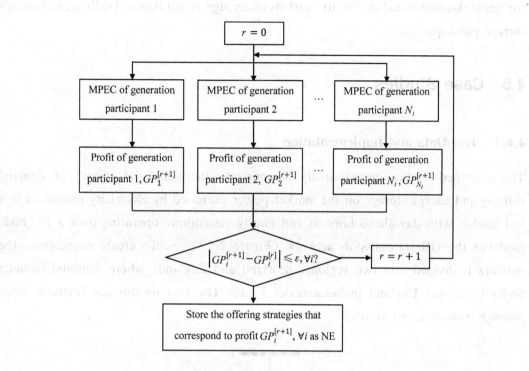

Figure 4.11 Flowchart of iterative diagonalization algorithm

diagonalization algorithm can be executed for multiple, randomly generated starting points[123] for the individual participants' offering decisions so as to increase the probability of finding multiple NE if NE exist. When multiple NE exist, authors in reference [55] allow selecting different solutions that meet certain pre-established criteria, such as achieving the maximum total profit for all strategic producers or obtaining the maximum social welfare for the market so as to induce market players to attain a particular equilibrium.

Furthermore, the iterative diagonalization approach is not generally guaranteed to converge to a market equilibrium, even if market equilibria exist, but if convergence is achieved, the resulting solution is a NE. Nevertheless, a market equilibrium has been reached within a relatively small number of iterations in every examined case study (Section 4.6). This finding, along with the focus of this work on investigating the impact of DS on producers' market power, sets a detailed analysis of existence, uniqueness, and convergence to a market equilibrium out of the scope of this paper.

To sum up, the methodology developed in this chapter aims at combining the iterative diagonalization algorithm with a MPEC approach to calculate supply function equilibria for network-constrained electricity markets under significant demand shifting and energy storage participation.

4.6 Case Studies

4.6.1 Test Data and Implementation

The examined studies quantitatively demonstrate the beneficial impact of demand shifting and energy storage on the market power exercised by electricity producers in a test market with day-ahead horizon and hourly resolution, operating over a 16-node model of the GB transmission network (Figure 4.12)[125]. To create congestion, the network is divided into two regions, Scotland and England, where Scotland includes nodes 1-6, and England includes nodes 7-16. The two regions are interconnected through transmission link (6,7).

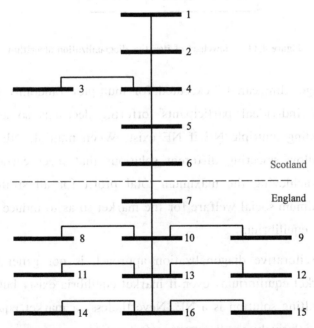

Figure 4.12 16-node model of GB transmission network

The market includes 7 generation participants, with their linear/quadratic cost coefficients, maximum output limits[127] and locations presented in Table 4.1. The increase in the value of the quadratic cost coefficient q_i^G as we move from the cheapest (base) unit 1 to the most expensive (peak) unit 7 expresses the increasing slope of the generation marginal cost curve — equal to $2 q_i^G$ according to equation (4.2) — as qualitatively illustrated in Figure 4.6. The distribution of the generation participants throughout the network reflects the situation in the Great Britain (GB) system, where the largest proportion of base units is located in Scotland while the largest proportion of mid and peak units is located in England (i.e. England is characterized by more expensive generation).

Table 4.1   Operational parameters and locations of generation participants

Generation participant i	1	2	3	4	5	6	7
$l_i^G / £ \cdot \text{MWh}^{-1}$	10	15	23	35	50	70	100
$q_i^G / £ \cdot \text{MW}^2 \text{h}^{-1}$	0.0001	0.0006	0.0014	0.0026	0.0042	0.0065	0.001
$g_i^{\max}/\text{MW}$	13,170	11,520	7,560	6,670	6,500	5,760	5,500
Node	3	9	5	11	6	15	16

The market also includes 13 demand participants, with their location and relative size (expressed as % of the total system demand and assuming that it remains identical for every time period) presented in Table 4.2. This table reflects the situation in the GB system, where the largest demand centres are located in England. As discussed in subsection 4.3.2, the coefficients of the demand side's benefit function and maximum baseline demand are time-specific parameters, following the daily pattern of consumers' activities. Figure 4.13 presents the assumed hourly values of maximum system baseline demand (which are derived based on a typical winter day demand profile of the GB system) and the hourly values of the linear benefit coefficient (which are assumed identical for every node)[126]. As discussed in subsection 4.3.1 and subsection 4.3.2, the quadratic generation cost and demand benefit functions are approximated by piecewise linear functions, consisting of 20 equal-size blocks each.

Table 4.2   Location and relative size of demand participants

Demand participant j	1	2	3	4	5	6	7	8	9	10	11	12	13
Node	1	2	4	5	6	7	8	9	11	12	14	15	16
Size/%	1.8	2.1	3.6	5.6	0.8	19.0	14.1	5.6	5.9	6.3	10.4	2.5	22.3

Different scenarios are examined regarding a) the time-shifting flexibility of the demand

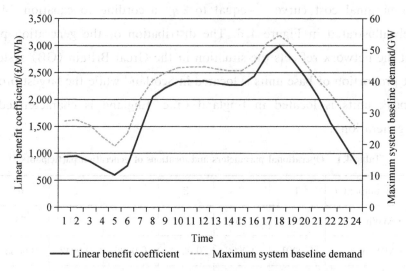

Figure 4.13 Hourly values of linear benefit coefficient and maximum system baseline demand

side, as expressed by parameter α_j, and b) the size of the energy storage unit in the market, as expressed by its capacity E^{cap} as a percentage β of the daily energy demand in the system. The assumed values of the rest of the energy storage operational parameters are given in Table 4.3.

Table 4.3 Operational parameters of energy storage participant

Parameter	E^{min}	E^{max}	E_0	s^{max}	η^c	η^d
Value	$0.2\,E^{cap}$	E^{cap}	$0.25\,E^{cap}$	$0.5\,E^{cap}/1\,h$	0.9	0.9

The developed equilibrium programming model was coded and solved using the optimization software FICO™ Xpress[127] on a computer with a 6-core 3.47 GHz Intel (R) Xeon(R) X5690 processor and 192 GB of RAM. The average computational time required for solving the MILP problem of a single strategic producer was around 20 s. A market equilibrium was reached in every examined scenario and the average number of required iterations was 24.

4.6.2 Impact of Demand Shifting and Energy Storage: Uncongested Network

This sub-section considers a case where the network capacity limits are neglected and therefore the network is not congested. For different demand shifting flexibility (assumed identical for every demand, i.e. $\alpha_j = \alpha$, $\forall j$) and energy storage capacity scenarios, two cases are compared: i) a case of perfectly competitive market (indicated by the superscript c in the remainder), where all producers behave competitively at all time periods, i.e. $k_{i,t}=1$, $\forall i$, $\forall t$, and ii) a case of imperfect, oligopolistic market (indicated by the superscript s in the remainder), where the offering strategies of the producers are determined based on the developed equilibrium model (Section 4.5). The demand and energy storage participants are assumed competitive, price-taking entities, revealing their actual economic and physical characteristics to the market (subsection 4.3.2 and subsection 4.3.3).

Market power can be measured by different indexes, which can be classified roughly as two different categories with reference to the cause or effect of market power that they employ for its quantification: the concentration indexes, based on the market shares of different players, and the comparison indexes, in which market performance in actual markets are compared with perfect competition. In this subsection, we resort to some comparison indexes to analyse and detect what will happen to prices, participants' economic surpluses and social welfare with attempt to capture the strategic aspects of competition in electricity markets[117-119].

We then investigate the logic of comparison indexes to compare the market outcomes between perfect and imperfect competition. The trend of these indexes along with the increase of demand shifting and energy storage flexibility helps to detect in which sense the market performance improves or worsens. All of them show an improvement when their values approach 0, because it means that the strategic oligopoly equilibrium is approaching the ideal benchmark represented by the perfect competition.

The *average Lerner index* (AveLI) (4.101) expresses the average increment of market prices driven by the exercise of market power. As qualitatively illustrated in subsection 4.4.2, demand shifting and energy storage reduce the price increment at peak periods and they increase it at off-peak periods, with the former reduction being prominently higher than the latter increase and resulting in an overall positive impact. This positive impact is illustrated in Figures 4.14 and 4.15, which demonstrate that AveLI is reduced with an increasing demand shifting flexibility and energy storage capacity.

$$AveLI = \underset{n,t}{\text{average}} \frac{\lambda_{n,t}^s - \lambda_{n,t}^c}{\lambda_{n,t}^s} \times 100\% \qquad (4.101)$$

This reduction of producers' ability to manipulate market prices has also an impact on their additional profit driven by the exercise of market power. Figures 4.16 and 4.17 present the aggregate increment of all producers' hourly profit for different demand shifting flexibility and energy storage capacity scenarios. Following the trend

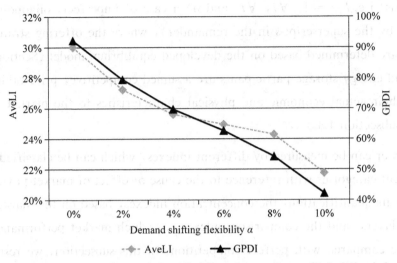

Figure 4.14 Average Lerner index (AveLI) and generation profit deviation index (GPDI) for different demand shifting flexibility scenarios

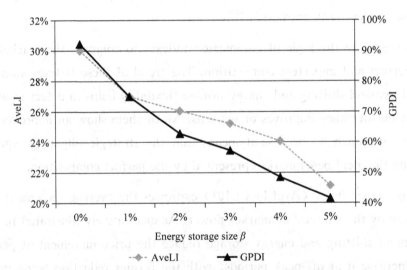

Figure 4.15 Average Lerner index (AveLI) and generation profit deviation index (GPDI) for different energy storage capacity scenarios

characterizing the price increments qualitatively illustrated in subsection 4.4.2, demand shifting and energy storage reduce the hourly profit increment during peak periods and increase it during off-peak periods, with the former reduction being significantly higher than the latter increase. As a result, the total profit increment driven by the exercise of market power is significantly reduced. This reduction is justified through the quantification of the *generation profit deviation index* (GPDI) (4.102). Figure 4.16 and Figure 4.17 demonstrate that GPDI is reduced with an increasing demand shifting flexibility and energy storage capacity, implying that the latter reduces the additional profit driven by the exercise of market power.

$$GPDI = \frac{\sum_{i,t} GP_{i,t}^s - \sum_{i,t} GP_{i,t}^c}{\sum_{i,t} GP_{i,t}^c} \times 100\% \quad (4.102)$$

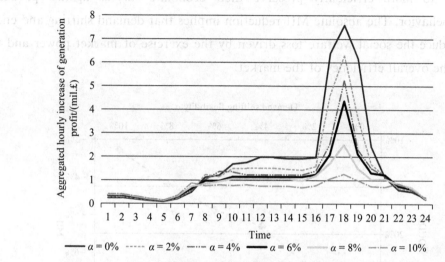

Figure 4.16 Hourly increase of generation profit driven by market power for different demand shifting flexibility scenarios

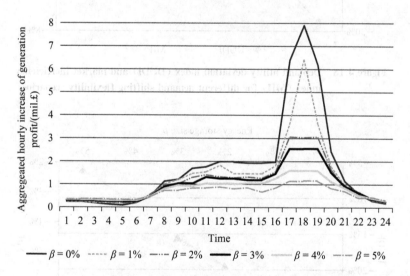

Figure 4.17 Hourly increase of generation profit driven by market power for different energy storage capacity scenarios

The reduction of the generation market power has also beneficial effects on demand utility and social welfare, which are justified by the quantification of the *demand utility deviation index* (DUDI) (4.103) and the *market inefficiency index* (MII) (4.104), respectively. Figures 4.18 and 4.19 demonstrate that (the absolute values of) DUDI and MII are both reduced with an increasing demand shifting flexibility and energy storage capacity. The absolute DUDI reduction implies that demand shifting and energy storage reduce the demand utility loss driven by the exercise of market power, and thus allow consumers to more efficiently preserve their economic surplus against producers' strategic behavior. The absolute MII reduction implies that demand shifting and energy storage reduce the social welfare loss driven by the exercise of market power and thus enhance the overall efficiency of the market.

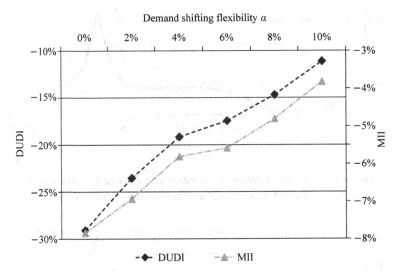

Figure 4.18 Demand utility deviation index (DUDI) and market inefficiency index (MII) for different demand shifting flexibility scenarios

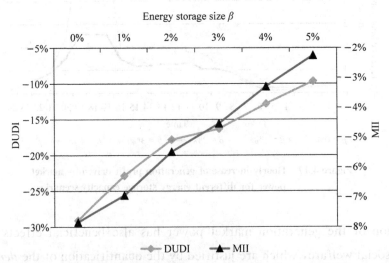

Figure 4.19 Demand utility deviation index (DUDI) and market inefficiency index (MII) for different energy storage capacity scenarios

$$DUDI = \frac{\sum_{j,t} DU^s_{j,t} - \sum_{j,t} DU^c_{j,t}}{\sum_{j,t} DU^c_{j,t}} \times 100\% \qquad (4.103)$$

$$MII = \frac{\sum_t SW^s_t - \sum_t SW^c_t}{\sum_t SW^c_t} \times 100\% \qquad (4.104)$$

To sum up, the MII assesses the market performance from the perspective of the social welfare, its decreasing trend along with the increase of demand shifting flexibility and energy storage capacity indicates that the overall market performance and economic efficiency are improved. Furthermore, strategic behaviour of the generation side result in a reallocation of surplus between generation and demand participants from what would be observed under perfect competition (or the maximum social welfare solution). In this context, the quantifications of GPDI and DUDI reveal such redistribution of surpluses. The decreasing trends of both quantities suggest that the surplus is shifted from the generation to the demand side. In other words, the strategic producers can game the market less to increase their surplus, whereas consumers are less affected by producers' market manipulation behaviours if they exhibits time-shifting flexibility, which consequently allows them to preserve more surplus.

4.6.3 Impact of Demand Shifting and Energy Storage: Congested Network

In this subsection the impact of network congestion and the location of demand shifting flexibility are investigated by examining the following cases:

U: Network capacity limits are neglected and therefore the network is uncongested, under both competitive and oligopolistic market settings. The demand side has no shifting flexibility.

U-DS-SC: The network is uncongested and demands in Scotland have shifting flexibility.

U-DS-EN: The network is uncongested and demands in England have shifting flexibility.

C: Network capacity limits are taken into account; in this case the line (6,7) connecting Scotland and England gets congested during some peak hours, under both competitive and oligopolistic market settings. The demand side has no shifting flexibility.

C-DS-SC: Line (6,7) is congested and demands in Scotland have shifting flexibility.

C-DS-EN: Line (6,7) is congested and demands in England have shifting flexibility.

Given that England is characterized by significantly higher demand than Scotland (Table 4.2), in order to provide a meaningful analysis regarding the impact of the location of demand shifting flexibility in the network, the overall extent of demand flexibility is assumed identical in cases U-DS-SC, U-DS-EN, C-DS-SC, and C-DS-EN, and equivalent to the $a=6\%$ scenario of subsection 4.6.2. It should be noted that case studies with energy storage instead of demand shifting have been executed, with the results exhibiting very similar trends to the above analysis. Therefore, these results are presented here for brevity reasons.

Figures 4.20 and 4.21 present the GPDI and DUDI corresponding to producers and demands in Scotland and England, and Figure 4.22 presents the MII, for each of the above cases. First of all, let us examine cases without demand shifting flexibility. When the network is uncongested, the locational prices are identical in the two areas, while congestion in line (6,7) yields a locational price differential between the two areas[46]. More specifically, during periods when the line is congested, England — characterized by more expensive generation and higher demand — exhibits higher price than the one observed in the uncongested case, while Scotland — characterized by cheaper generation and lower demand — exhibits lower price than the one observed in the uncongested case. Producers' market power is enhanced at higher price levels due to the increasing slope of the strategic offering curve (subsection 4.4.2). Therefore, congestion increases the GPDI and the absolute value of DUDI corresponding to producers and demands in England, while it reduces the GPDI and the absolute value of DUDI corresponding to producers and demands in Scotland, as observed in Figures 4.20 and 4.21. In other words, congestion creates a more favourable setting for producers in England and demands in Scotland, and a less favourable setting for producers in Scotland and demands in England. The overall impact of congestion on the efficiency of the market is negative (as justified by the increase of the absolute value of MII in Figure 4.22), since the negative impact on producers' market power in the higher-priced area (England) dominates the positive impact in the lower-priced area (Scotland). These findings verify the conclusions of previous work [48], [50], [55], [115], [117] that network congestion favours market power exercise by strategically-located producers and aggravates the overall impacts of market power exercise.

Let us now examine the impact of introducing demand shifting flexibility in the market. When the network is uncongested, producers' market power is reduced (subsection 4.4.2), resulting in a reduction of the MII as well as the GPDI and the absolute value of

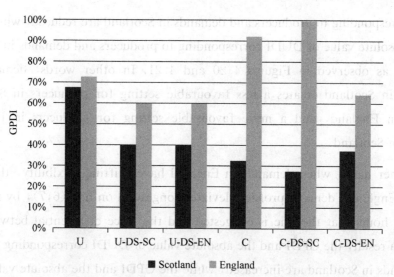

Figure 4.20 Generation profit deviation index (GPDI) corresponding to producers in Scotland and England for each of the examined cases

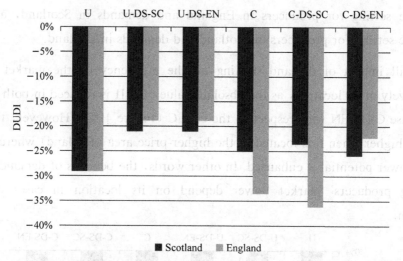

Figure 4.21 Demand utility deviation index (DUDI) corresponding to demands in Scotland and England for each of the examined cases

DUDI for producers and demands in both areas. Furthermore, the location of demand shifting flexibility does not have an impact on the market outcome and thus cases U-DS-SC and U-DS-EN exhibit the same indexes' values (Figures 4.20 and 4.21).

When the network is congested though, the location of demand shifting flexibility affects the market outcome significantly. When demands in Scotland have shifting flexibility, the flattening effect on Scotland's demand profile aggravates congestion on line (6,7), by increasing the number of hours that the line is congested and the price differential between the two areas. As a result, the GPDI and the absolute value of

DUDI corresponding to producers and demands in Scotland are reduced, while the GPDI and the absolute value of DUDI corresponding to producers and demands in England are increased, as observed in Figures 4.20 and 4.21. In other words, demand shifting flexibility in Scotland creates a less favourable setting for producers in Scotland and demands in England, and a more favourable setting for producers in England and demands in Scotland.

On the other hand, when demands in England have shifting flexibility, the flattening effect on England's demand profile alleviates congestion on line (6,7), by reducing the number of hours that the line is congested and the price differential between the two areas. As a result, the GPDI and the absolute value of DUDI corresponding to producers and demands in Scotland are increased, while the GPDI and the absolute value of DUDI corresponding to producers and demands in England are reduced, as observed in Figures 4.20 and 4.21. In other words, demand shifting flexibility in England creates a less favourable setting for producers in England and demands in Scotland, and a more favourable setting for producers in Scotland and demands in England.

The overall impact of demand shifting on the efficiency of the market is positive irrespectively of its location, as the absolute value of MII is reduced in both case C-DS-SC and case C-DS-EN with respect to the case C (Figure 4.22). However, this positive impact is higher when it is located in the higher-price area (England) where producers' market power potential is enhanced. In other words, the benefits of demand shifting in mitigating producers' market power depend on its location in cases of network congestion.

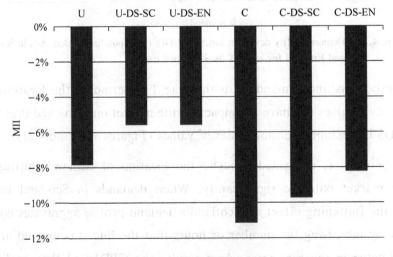

Figure 4.22 Market inefficiency index (MII) for each of the examined cases

4.7 Conclusions

The role of the demand side in imperfect electricity markets has been previously investigated in terms of the effect of its own-price elasticity on electricity producers' ability to exercise market power. However, consumers' flexibility regarding electricity use cannot be fully captured through the concept of own-price elasticity. Instead of simply avoiding using their loads at high price levels, consumers are more likely to shift the operation of their loads from periods of higher prices to periods of lower prices. This shift of energy demand from high- to low-priced periods drives a demand profile flattening effect. A similar effect is driven by energy storage technologies which are charged during periods of lower prices and are discharged during periods of higher prices.

This chapter has provided for the first time theoretical and quantitative evidence of the beneficial impact of flexibility enabled by demand shifting and energy storage in generation market power mitigation. Theoretical explanation of this impact has been presented through a simple price-quantity graph in a simplified two-period market. This graph has demonstrated that demand shifting and energy storage reduce the increase in market prices driven by the exercise of market power at peak periods and increase it at off-peak periods, with the former reduction dominating the latter increase and resulting in an overall positive impact.

Quantitative analysis has been supported by a multi-period equilibrium programming model of the oligopolistic market setting. The decision-making process of each strategic generation participant is modelled through a bi-level optimization problem. The upper level represents the profit maximization objective of the participant, while the lower level represents the market clearing process, accounting for the time-coupling operational constraints of demand shifting and energy storage, and the network constraints. This bi-level problem is solved after converting it to a MPEC and linearizing the latter through suitable techniques. The oligopolistic market equilibria resulting from the interaction of multiple independent generation participants are determined by employing an iterative diagonalization method.

Case studies with the developed model on a test market reflecting the general generation and demand characteristics of the GB system have quantitatively demonstrated the benefits of demand shifting and energy storage in limiting producers' market power, by employing relevant indexes from the literature. In cases without network congestion, the location of demand shifting and energy storage does not have an impact on producers' market power exercise, but an increasing demand shifting and energy storage flexibility is shown to i) reduce strategic producers' ability to manipulate market prices, ii) reduce strategic producers' additional profit driven by the exercise of market power, iii) allow consumers to more efficiently preserve their economic surplus against producers' strategic behaviours, and iv) reduce the social welfare loss and thus enhance the overall efficiency of the market. In cases with network congestion, demand shifting and energy storage flexibility still has an overall positive impact on market efficiency, but the extent of this benefit as well as the impact on producers and demands at different areas depends on the location of demand shifting and energy storage in the network.

Nomenclature

Chapter 4 and Chapter 5 share this nomenclature.

Indices and Sets

t	Index of time periods.
i	Index of generation participants.
I_n	Set of generation participants connected to node n.
j	Index of demand participants.
J_n	Set of demand participants connected to node n.
b	Index of generation blocks.
c	Index of demand blocks.
n, m	Index of nodes.
M_n	Set of nodes connected to node n.

Parameters

N_T	Number of time periods of the horizon.
N_I	Number of generation participants.
l_i^G	Linear cost coefficient of generation participant i (£/MWh).
q_i^G	Quadratic cost coefficient of generation participant i (£/MW²h).
g_i^{max}	Maximum power output limit of generation participant i (MW).
$\lambda_{i,b}^G$	Marginal cost of block b of generation participant i (£/MWh).
$g_{i,b}^{max}$	Maximum power output limit of block b of generation participant i (MW).
$l_{j,t}^D$	Linear benefit coefficient of demand participant j at time period t (£/MWh).
$q_{j,t}^D$	Quadratic benefit coefficient of demand participant j at time period t

	($£/MW^2h$).
$d_{j,t}^{\max}$	Maximum power demand of demand participant j at time period t (MW).
$\lambda_{j,t,c}^{D}$	Marginal benefit of block c of demand participant j at time period t ($£/MWh$).
$d_{j,t,c}^{\max}$	Maximum power demand of block c of demand participant j at time period t (MW).
α_j	Load shifting of demand participant j.
$s^{\max}$	Power rating of energy storage (MW).
E^{cap}	Energy capacity of energy storage (MWh).
$E^{\min}$	Minimum energy limit of energy storage (MWh).
$E^{\max}$	Maximum energy limit of energy storage (MWh).
E_0	Initial energy level in energy storage (MWh).
η^c	Charging efficiency of energy storage.
η^d	Discharging efficiency of energy storage.
n_s	Node to which energy storage is connected.
$F_{n,m}^{\max}$	Capacity of transmission line (n, m) (MW).
$x_{n,m}$	Reactance of transmission line (n, m) (p.u.).

Variables

$k_{i,t}$	Strategic offer variable of generation participant i at time period t.
$k_{i-,t}$	Strategic offer variable of generation participants other than i at time period t.
$g_{i,t,b}$	Power output of block b of generation participant i at time period t (MW).
$d_{j,t,c}$	Power demand of block c of demand participant j at time period t (MW).
$d_{j,t}^{\text{sh}}$	Change of power demand of demand participant j at time period t due to load shifting (MW).
$d_{j,t}'$	Power demand after load shifting of demand participant j at time period t (MW).
s_t^c	Charging power of energy storage at time period t (MW).
s_t^d	Discharging power of energy storage at time period t (MW).
E_t	Energy level in energy storage at the end of time period t (MW).
$\theta_{n,t}$	Voltage angle of node n at time period t (rad).
$\lambda_{n,t}$	Locational marginal price at node n at time period t ($£/MWh$).
$\mu, \nu, \xi, \pi, \rho, \sigma, \varphi, \chi, \psi, \beta, \gamma, \delta$	
	Dual variables corresponding to the lower-level constraints. See subsection 4.5.1 and subsection 5.5.1 for details.
ω	Auxiliary variables correspond to the linearization of complementarity conditions.

Functions

$C_{i,t,b}$	Cost of block b of generation participant i at time period t ($£/h$).

$C'_{i,t,b}$ Marginal cost of block b of generation participant i at time period t (£/MWh).

$SC_{i,t,b}$ Strategic cost of block b of generation participant i at time period t (£/MWh).

$SC'_{i,t,b}$ Strategic marginal cost of block b of generation participant i at time period t (£/MWh).

$GP_{i,t}$ Profit of generation participant i at time period t (£/h).

$B_{j,t,c}$ Benefit of block c of demand participant j at time period t (£/h).

$B'_{j,t,c}$ Marginal benefit of block c of demand participant j at time period t (£/MWh).

$DU_{j,t}$ Utility of demand participant j at time period t (£/h).

$DP_{j,t}$ Payment of demand participant j at time period t (£/h).

SP_t Profit of energy storage at time period t (£/h).

SW_t Social welfare at time period t (£/h).

Chapter 5
Investigating the Exercise of Market Power by Strategic Flexible Demand and Energy Storage in Imperfect Electricity Markets

5.1 Introduction

The deregulated electricity markets are better described in terms of imperfect rather than perfect competition (Section 4.1). In this setting, market participants do not necessarily act as price takers. Participants of large size and/or strategically located in the transmission network are able to exercise market power, i.e. influence the electricity prices and increase their profits beyond the competitive equilibrium levels, through strategic bids and offers.

The increasing penetration of demand response and energy storage technologies in power systems, driven by the emerging Smart Grid paradigm[56], has recently attracted significant research interest in exploring the role of these technologies in imperfect electricity markets. Chapter 4 has qualitatively and quantitatively demonstrated that the flexibility enabled by demand shifting and energy storage reduces the extent of exercised market power by the generation side at peak periods and increases it at off-peak periods, with the former reduction dominating the latter increase and resulting in an overall positive impact.

As discussed in Section 1.3, the second aspect of this research regarding imperfect electricity markets lies in exploring the ability of the demand side and energy storage to exercise market power. Regarding the former, previous work [57]—[58] has demonstrated that large consumers can strategically increase their surplus by underbidding their actual marginal benefit (subsection 5.2.1). However, given the envisaged enhancement of demand time-shifting flexibility (subsection 4.3.2), consumers' ability to exercise market power through revealing less flexibility potential to the market is not discussed in the literature. Concerning the latter, previous work in references [59]—[62] has demonstrated the ability of large ES units to exercise market power by withholding their capacity, leading to additional ES profits but loss of social welfare. However, the modelling approaches adopted in references [59]—[61] exhibit certain limitations, while the dependency of the extent of exercised market power on ES and system parameters are not properly analysed in previous studies (subsection 5.2.2).

This chapter aims at addressing the above challenges. Firstly, the market potential of the demand side and energy storage is qualitatively analysed through a price-quantity graph in a simplified two-period market. Secondly, a multi-period game-theoretic model (Section 5.5) is developed for optimizing capacity withholding strategies of energy storage. Case studies with the developed model on a test market with day-ahead horizon and hourly resolution, operating over a 16-node transmission network quantitatively analyse the extent of storage capacity withholding and its impact on its profit and social welfare at different time periods and for different scenarios regarding: i) the size of energy storage (in terms of its power rating and energy capacity), ii) the characteristics of the demand side (in terms of its profile shape and price elasticity), iii) the characteristics of the generation side (in terms of the installed capacity and output profile shape of wind generation), and iv) the location of energy storage in the presence of network congestion, resorting to relevant market power metrics from the literature.

These qualitative and quantitative insights are crucial for both owners and potential investors in energy storage technologies, in terms of devising suitable strategies in electricity markets, as well as regulators and policy makers, in terms of analysing the role and impacts of energy storage in the deregulated electricity sector.

This chapter is organized as follows. Section 5.2 conducts a detailed literature review regarding modelling strategic behaviour of the demand side and optimal operation of energy storage in deregulated electricity markets. Section 5.3 outlines models of generation, demand, and strategic storage participants. Section 5.4 provides qualitative analysis of the market power potential of strategic demand side and energy storage. Section 5.5 formulates the bi-level optimization model expressing the decision-making of strategic energy storage and derives the equivalent MPEC and MILP formulations. Case studies for different scenarios are presented in Section 5.6. Section 5.7 discusses conclusions of this chapter.

5.2 Literature Review

5.2.1 Modelling Strategic Behaviour of Demand Participants

The emerging Smart Grid paradigm has enabled an increasing level of demand response, facilitates the strategic behaviour of the demand side. In this context, recent work has demonstrated that large consumers can strategically increase their surplus by submitting bids lower than their actual marginal benefit[57-58]. In reference [57], a stochastic complementarity model is developed that allows large consumers to derive optimal bidding strategies to manipulate market prices to their own benefit in an imperfect electricity market with producers' offer uncertainties. In reference [58], the same authors explore the effects of allowing large consumers to exercise market power in both energy and reserve markets with significant penetration of wind generation.

At the same time, the ability of the demand side to redistribute its energy requirements across time (subsection 4.3.2) will be enhanced with the increasing penetration of various flexible demand technologies, such as smart-charging electric vehicles and smart domestic appliances, and advanced metering, control and communication technologies in consumers' premises. In this context, the possibility of the demand side to exercise market power through revealing less flexibility potential to the market is not discussed in the literature. This chapter aims to fill this knowledge gap (subsection 5.4.1).

5.2.2 Modelling Strategic Behaviour of Energy Storage Participants

Approaches examined in the literature regarding modelling optimal operation of energy storage systems can be broadly divided into two broad categories. In the first category[19], [28-29], [84-85], storage units are treated as system assets utilized by the system operator in a centrally administered market (subsection 2.3.1). In this context, energy storage could promise significant system benefits including improving system reliability, supporting system balancing, provision of ancillary services and facilitating high renewable energy penetration, etc. In this category, the operation of the storage unit is optimized from the system perspective so as to quantify the values of energy storage to the electricity system. In other words, the storage unit is not considered as an independent profit-driven entity, and thus its profitability is not the focus.

In the second category, storage units are operated by profit-seeking merchants[59-62], [80-81].

These studies can further be divided into two groups depending on whether storage could influence market prices. In the first group, the storage units are treated as price-takers (i.e. they are assumed small-scale so that their operation has negligible influence on the market price) in the market (subsection 2.3.2). In this context, the operation of storage is generally optimized under the assumption that the future market prices are known in advance and are considered as exogenous parameters. Hence, the storage operator would utilize this forecasted price information to optimize its operation decisions. In reference [80], a storage unit is owned by a wind farm and its operation decisions are established according to the wind farm's power production estimates in order to maximize the owner's profit. In reference [81], a storage unit is operated by an independent utility to maximize its profit obtained in both energy and ancillary markets.

The limitation of all these studies is the assumption that storage operations have no impact on market price formation. However, storage purchasing and selling activities will change the net demand and supply, and consequently the market price. Thus, the static price assumption could overstate the arbitrage value of storage, due to the ignorance of the effect of storage in suppressing market price differentials[124]. As a consequence, a profit-maximizing storage operator could *underuse* storage compared to the welfare optimum. To this end, the second group relaxes the competitive assumption and treats energy storage units as price-makers (i.e. they are assumed large-scale and their operation can manipulate market prices) in the market. In such imperfect market setting, previous work [59]—[62] has demonstrated their ability to exercise market power by withholding their capacity, leading to additional storage profits but loss of social welfare. In reference [59], a Cournot oligopoly equilibrium model is adopted to analyse the potential for exerting market power that is associated with the storage operations. The impact of strategic storage on market prices is modelled with an inversed demand function. The authors in reference [60] employ analytical calculation of market equilibria in a simplified market example. The model is limited to only examining two time periods representing off-peak and on-peak demand. The result shows that strategic storage operation results in reduced social welfare compared to a perfectly competitive market, and this effect is more significant if generators in the same market behave strategically. In reference [61], the same authors explore the effects of strategic storage operation and ownership on market welfare in the same two-period market as in reference [60]. The analysis shows that strategic storage dispatch always results in suboptimal welfare regardless of ownership. When the ownership belongs to generators and merchant storage operators, the storage capacity tends to be underused to prevent drop in price (at the peak period) that would reduce the profits of these parties. On the contrary, consumers tend to overuse the storage as it is their interest to drive the price down. The operation of a number of large, price-maker, and locational dispersed energy storage units in a network-constrained electricity market is examined in reference [62]. The energy storage units are assumed to be investor-owned and independently-operated, seeking to maximize their total profits. The impact of storage parameters on its strategic operation is investigated.

The modelling approaches adopted in references [59]—[61] however exhibit certain limitations. In reference [59], the parameters of the employed inverse demand function (expressing the relation between the price and net demand) are determined through exogenous data and therefore cannot accurately capture the impacts of the market's generation and demand characteristics on price formation. It should also be noted that the inverse demand function in reference [59] is exponential, meaning that the formulated optimal decision-making problem of strategic storage is nonlinear, and any solution obtained by commercial solvers is not guaranteed to be globally optimal. The complex analytical calculation of market equilibria conducted in references [60]—[61] cannot be easily extended to realistic market models involving a much larger number of clearing periods. Furthermore, none of the above papers analyse properly the dependency of the extent of market power exercised by strategic storage (i.e. the additional storage profit and the social welfare loss driven by the exercise of capacity withholding) on its size and the characteristics of the generation and demand sides of the market. In reference [62], although the impact of storage parameters on its strategic behaviour is analysed, the profitability of the storage facility under strategic and competitive behaviour is not compared, and therefore this work does not aim to quantify the extent of market power exercised by the strategic storage. However, such comparison provides crucial insights of market structures' (i.e. with perfect/imperfect competition) impacts on storage units' operation decisions. References [59]—[61] neglect network constraints and are therefore unable to investigate the impact of the location of storage on its market power potential.

This chapter tackles the above challenges. The decision-making process of strategic storage is modelled through a bi-level optimization problem, where the upper level determines the optimal extent of capacity withholding by strategic storage at different time periods, maximizing the storage profit, while the lower level represents endogenously the market clearing process (capturing the impact of strategic storage operation on the market clearing outcomes), accounting for network constraints. This problem is converted to a MPEC, by replacing the lower level problem by its equivalent KKT optimality conditions. Suitable linearization technique is proposed to convert this non-linear MPEC to a MILP which can be solved using available commercial solvers. Case studies with the proposed model quantitatively analyse the extent of capacity withholding and its impact on storage profit and social welfare for different scenarios regarding the size of storage, the characteristics of the generation and demand sides of the market, as well as the location of storage in case the network congestion is taken into account.

5.3 Modelling Market Participants

5.3.1 Generation Participants

Following the model of the generation participants presented in subsection 4.3.1, it is assumed that each generation participant i owns a single unit. The quadratic cost function of which is approximated by a piecewise linear cost function (5.1), consisting of a number of blocks[90]. The associated stepwise marginal cost function and output limits of each block b are expressed as:

$$C_{i,t,b}(g_{i,t,b}) = \lambda_{i,b}^G g_{i,t,b} \tag{5.1}$$

$$C'_{i,t,b}(g_{i,t,b}) = \lambda_{i,b}^G \tag{5.2}$$

$$0 \leqslant g_{i,t,b} \leqslant g_{i,b}^{\max}, \forall t \tag{5.3}$$

As comprehensively explored in Chapter 4, strategic generation participants can exercise market power by either submitting offers higher than their actual marginal costs (i.e. economic withholding) or offering less than their actual generation capacity to the market (i.e. capacity withholding). However, since the focus of this chapter is on the market power potential of energy storage, generation participants are assumed competitive, price-taking entities, revealing their actual economic and physical characteristics to the market.

5.3.2 Demand Participants

Following the model of the demand participants presented in subsection 4.3.2, the quadratic benefit function of which is approximated by a piecewise linear benefit function (5.4), consisting of a number of blocks. The associated stepwise marginal benefit function and demand limits of each block c are expressed as:

$$B_{j,t,c}(d_{j,t,c}) = \lambda^{D}_{j,t,c} d_{j,t,c} \tag{5.4}$$

$$B'_{j,t,c}(d_{j,t,c}) = \lambda^{D}_{j,t,c} \tag{5.5}$$

$$0 \leqslant d_{j,t,c} \leqslant d^{\max}_{j,t,c}, \; \forall t \tag{5.6}$$

The time-shifting flexibility of the demand participant j is expressed by constraints (5.7)—(5.9). Constraint (5.7) expresses that the net demand of participant j at time period t encompasses two components: the baseline level $\sum_c d_{j,t,c}$ and change of the demand level $d^{sh}_{j,t}$ due to load shifting (taking negative values when demand is moved away from t and positive values when demand is moved towards t). Constraint (5.8) ensures that demand shifting is energy neutral within the examined time horizon i.e. the total size of demand reductions is equal to the total size of demand increases, assuming that demand shifting does not involve energy gains or losses. Constraint (5.9) expresses the limits of demand change at each time period due to load shifting as a ratio α_j ($0 \leqslant \alpha \leqslant 1$) of the baseline demand; $\alpha_j = 0$ implies that the demand participant j does not exhibit any time-shifting flexibility, while $\alpha_j = 1$ implies that the whole demand of participant j can be shifted in time.

$$d'_{j,t} = \sum_c d_{j,t,c} + d^{sh}_{j,t}, \; \forall t \tag{5.7}$$

$$\sum_t d^{sh}_{j,t} = 0 \tag{5.8}$$

$$-\alpha_j \times \sum_c d_{j,t,c} \leqslant d^{sh}_{j,t} \leqslant \alpha_j \times \sum_c d_{j,t,c}, \; \forall t \tag{5.9}$$

5.3.3 Strategic Energy Storage

As mentioned in Section 2.2, the storage unit is considered as an independent market player, which is neither part of a storage-solar[68] nor storage-wind[69] coalition, and is neither part of the portfolio of a generation company[59], nor owned by different market players[60]. While those are viable cases, modelling them is beyond the scope of this chapter. Following the model presented in subsection 4.3.3, the operational characteristics of an energy storage unit are expressed by (5.10)—(5.14). Constraint (5.10) expresses the energy balance in the storage unit including charging and discharging losses. Constraint (5.11) corresponds to its maximum depth of discharge and state of charge ratings. Constraints (5.12)—(5.13) represent its charging and discharging power limits. For the sake of simplicity, the storage energy content at the start and the end of the market's temporal horizon are assumed equal as equation, ensuring energy conservation.

$$E_t = E_{t-1} + \eta^c s_t^c - s_t^d / \eta^d, \quad \forall t \tag{5.10}$$

$$E^{\min} \leqslant E_t \leqslant E^{\max}, \quad \forall t \tag{5.11}$$

$$0 \leqslant s_t^c \leqslant s^{\max}, \quad \forall t \tag{5.12}$$

$$0 \leqslant s_t^d \leqslant s^{\max}, \quad \forall t \tag{5.13}$$

$$E_0 = E_{N_T} \tag{5.14}$$

The profit obtained by the energy storage unit is expressed as the difference between the revenue it receives from selling the energy by discharging, and the cost it incurs from purchasing the energy by charging.

$$SP(\lambda_{n_s,t}, s_t^c, s_t^d) = \sum_t \lambda_{n_s,t} (s_t^d - s_t^c) \tag{5.15}$$

5.4 Qualitative Analysis of Demand Side and Energy Storage Market Power Capability

5.4.1 Market Power Potential of the Demand Side

- *Revealing lower marginal benefit to the market*

If a demand participant behaves in a competitive (non-strategic) manner, a comparatively higher demand of that participant is supplied, but at the cost of clearing the market at comparatively higher prices. Driven by this analysis, a strategic demand participant may increase its surplus by submitting bids lower than its actual marginal benefit and hope this will decrease the market clearing price[57].

As illustrated in Figure 5.1, the strategic action has a positive impact on demand utility since it decreases the market price at t (from λ_t^c to λ_t^s), and thus reduces the demand payment. However, it also has a negative effect on demand utility since it reduces the demand supplied to the consumers, and thus reduces the benefit obtained by the latter. Consequently, the demand participant should devise suitable strategy that optimally balances the trade-off between lower market price and lower quantity of demand supplied.

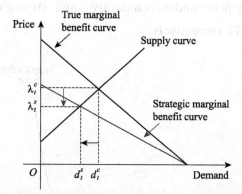

Figure 5.1 Illustration of the ability of demand participant to exercise market power through marginal benefit withholding

- *Revealing lower flexibility potential to the market*

As mentioned in subsection 1.3.2, the ability of demand participant to exercise market power by revealing lower flexibility potential to the market is qualitatively analysed in this subsection, through a price-quantity graph in a simplified market representation involving two periods (peak and off-peak) and no network constraints. The solid curve represents in a simplified fashion the aggregate offer curve of the generation side — characterized by increasing slope (subsection 4.4.2) — and vertical dashed lines represent the system demand in different cases; the intersection of the marginal cost curve with a vertical line determines the market clearing price in the respective case. Superscripts c and s denote competitive (i.e. non-strategic) and strategic behaviour of the demand participant respectively, while subscripts 1 and 2 denote off-peak and peak time periods respectively.

Two cases concerning the behaviour of the demand participant in the market are compared i) competitive behaviour, where the participant reveals its actual time-shifting flexibility to the market ($\alpha = \alpha^c$, where α^c represents the maximum limit of load shifting), and ii) strategic behaviour, where the participant reveals less time-shifting flexibility to the market ($\alpha < \alpha^c$).

As shown in Figure 5.2, the competitive demand behaviour derives a (net) demand profile flattening effect by reducing demand at the peak period from Q_2 to Q_2^c and increasing it at the off-peak period from Q_1 to Q_1^c. By acting strategically and revealing less time-shifting flexibility to the market, the demand participant limits this flattening effect on the system demand profile, since it i) increases less the demand at the off-peak period (from Q_1 to Q_1^s) and ii) reduces less the demand at the peak period (from Q_2 to

133

Q_2^s). The demand payment under competitive and strategic behaviour is given by equations (51.6)—(5.17) respectively.

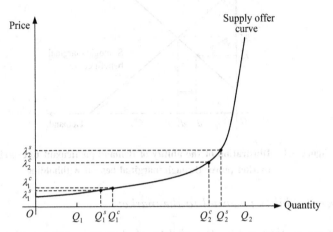

Figure 5.2 Illustration of market power potential of demand side through revealing less time-shifting flexibility to the market

$$DP^c = \lambda_1^c Q_1^c + \lambda_2^c Q_2^c \qquad (5.16)$$

$$DP^s = \lambda_1^s Q_1^s + \lambda_2^s Q_2^s \qquad (5.17)$$

As illustrated in Figure 5.2, this strategic action has a positive effect on demand payment, since it reduces the price and net demand at the off-peak period from λ_1^c to λ_1^s and from Q_1^c to Q_1^s, respectively. This leads to a demand payment decrement of $\lambda_1^c Q_1^c - \lambda_1^s Q_1^s$. However, it also has a negative effect on demand payment, since it increases the price and net demand at the peak period from λ_2^c to λ_2^s and from Q_2^c to Q_2^s, respectively. This leads to a demand payment increment of $\lambda_2^s Q_2^s - \lambda_2^c Q_2^c$. Due to the larger slope of the marginal cost of generation (subsection 4.3.2), the payment increment at the peak period is more prominent than its decrement at the off-peak period (i.e. $\lambda_2^s Q_2^s - \lambda_2^c Q_2^c > \lambda_1^c Q_1^c - \lambda_1^s Q_1^s$), resulting in an overall increase in consumers' electricity bills. As a result, consumers will find it financially unattractive to behave in such a strategic fashion.

Driven by the above observation, demand participants could potentially exercise market power by promising higher time-shifting flexibility to the market. As will be analysed in subsection 6.2.2, this behaviour contributes to an overall reduction of demand payment. However, this necessitates the employment of a *secondary market*[46] where the participants need to balance their position, as their demand may be supplied at a lower or higher quantity in the primary market under such strategic behaviour. In this context, future work will focus on devising optimal strategies for large consumers that participate in both primary and secondary markets.

5.4.2 Market Power Potential of Energy Storage

The concept of economic withholding (subsection 4.3.1 and subsection 5.4.1) does not generally apply to the case of energy storage, since the latter does not exhibit an explicit cost or benefit function. However, energy storage can exercise market power through capacity withholding i.e. by reporting lower than its actual power capacity to the market. In order to quantitatively capture the extent of capacity withholding at different time periods, the offered charging and discharging power capacity is expressed by constraints (5.18)—(5.19) respectively. The value of the decision variable $0 \leqslant k_t \leqslant 1$ represents the capacity withholding strategy of energy storage at period t. If $k_t = 0$, energy storage behaves competitively and reveals its actual power capacity to the market at t. If $k_t > 0$, energy storage behaves strategically and reports lower than its actual power capacity to the market at t.

$$0 \leqslant s_t^c \leqslant (1 - k_t) s^{\max}, \forall t \tag{5.18}$$

$$0 \leqslant s_t^d \leqslant (1 - k_t) s^{\max}, \forall t \tag{5.19}$$

Figure 5.3 illustrates, in a price-quantity graph, the ability of energy storage to exercise market power through capacity withholding in the same market representation as in the previous subsection.

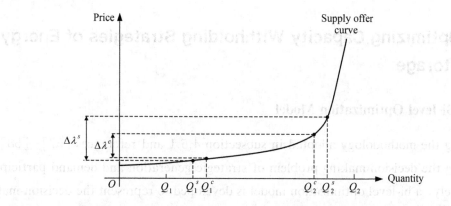

Figure 5.3 Illustration of market power exercise by energy storage through capacity withholding

As well established in relevant literature and discussed in subsection 4.4.2, the operation of competitive energy storage in the electricity markets flattens the (net) demand profile by i) charging and thus increasing demand at off-peak period from Q_1 to Q_1^c and ii) discharging and thus reducing demand at peak period from Q_2 to Q_2^c. By acting strategically and offering less power capacity to the market, energy storage limits its flattening effect on the system demand profile, since it i) charges less and thus increases less the demand at off-peak period (from Q_1 to Q_1^s) and ii) discharges less and thus reduces less the demand at peak period (from Q_2 to Q_2^s).

Although this strategic action has a negative effect on social welfare, it can be beneficial for energy storage. In the example of Figure 5.3, assuming a lossless energy storage for the sake of simplicity (implying that its charging and discharging quantities are equal, i.e. $Q_1^c - Q_1 = Q_2 - Q_2^c$ and $Q_1^s - Q_1 = Q_2 - Q_2^s$), its profit under competitive and strategic behaviour is given by equations (5.20)—(5.21) respectively.

$$SP^c = \lambda_2^c(Q_2 - Q_2^c) - \lambda_1^c(Q_1^c - Q_1) = \Delta\lambda^c(Q_2 - Q_2^c) \tag{5.20}$$

$$SP^s = \lambda_2^s(Q_2 - Q_2^s) - \lambda_1^s(Q_1^s - Q_1) = \Delta\lambda^s(Q_2 - Q_2^s) \tag{5.21}$$

This strategic action has a positive effect on energy storage profit, since it increases the price differential between peak and off-peak periods from $\Delta\lambda^c$ to $\Delta\lambda^s$, as demonstrated in Figure 5.3. However, it also has a negative effect on energy storage profit, since it reduces the volume of energy sold by storage ($Q_2 - Q_2^s < Q_2 - Q_2^c$). This mixed effect of capacity withholding on the profit of energy storage means that the latter needs to optimize its capacity withholding strategy k_t in order to maximize its profit in the market. The proposed formulation of this optimization problem is presented in the next section.

5.5 Optimizing Capacity Withholding Strategies of Energy Storage

5.5.1 Bi-level Optimization Model

Following the methodology adopted in subsection 4.5.1 and references [57]—[58] for modelling the decision-making problem of strategic generation and demand participants respectively, a bi-level optimization model is developed to represent the decision-making of strategic energy storage. As illustrated in Figure 5.4, the bi-level model is constitutive of an upper level (UL) problem as well as a lower level (LL) problem. In the UL, the energy storage behaves strategically through the capacity withholding decision made in the UL to maximize its profit. The UL problem is subject to the LL problem that represents endogenously the market clearing process, maximizing the social welfare, accounting for the *perceived* (since the energy storage may not report its actual power capacity (subsection 5.4.2) to the market) time-coupling operational characteristics of energy storage, and employing a DCOPF model to account for the effect of the network.

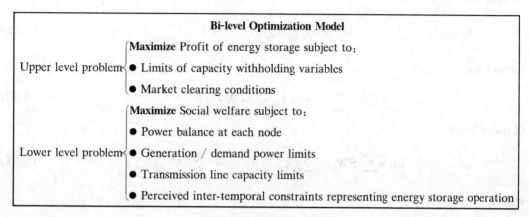

Figure 5.4 Bi-level structure of the decision-making problem of the strategic energy storage

Energy storage optimizes its offering strategy in the UL while receiving feedback (i.e. LMPs and storage charging/discharging dispatch) from the LL regarding how its capacity withholding action affects the market outcome. Therefore, the capacity withholding decisions are variables in the UL problem while considered as parameters in the LL problem.

The UL and LL problems are coupled as illustrated in Figure 5.5. The capacity withholding strategies determined by the upper level problem affect the constraints (i.e. the storage charging/discharging power limits) of the LL problem, and thus affect the market outcomes. The LMPs and storage charging/discharging dispatch determined by the lower level problem influence the objective function (i.e. the profit of the energy storage participant) of the UL problem.

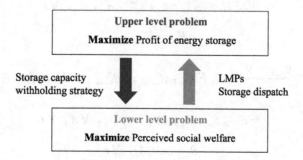

Figure 5.5 Interrelation between the upper level and lower level problems

The formulation of the proposed bi-level model of the strategic energy storage participant is given by (5.22)—(5.35) below. The dual variables associated with the LL problem (5.24)—

(5.35) are indicated at their corresponding constraints following a colon.

(Upper level)

$$\max_{\{k_t\}} \sum_t \lambda_{ns,t}(s_t^d - s_t^c) \quad (5.22)$$

subject to:

$$0 \leqslant k_t \leqslant 1, \ \forall t \quad (5.23)$$

(Lower level)

$$\max_{V^{\text{LL-P}}} \sum_t SW_t = \sum_{t,c} \lambda_{j,t,c}^D d_{j,t,c} - \sum_{i,t,b} \lambda_{i,b}^G g_{i,t,b} \quad (5.24)$$

where:

$$V^{\text{LL-P}} = \{g_{i,t,b}, d_{j,t,c}, d_{j,t}^{\text{sh}}, s_t^c, s_t^d, E_t, \theta_{n,t}\} \quad (5.25)$$

subject to:

$$\sum_{(j \in J_n),c} d_{j,t,c} + (s_t^c - s_t^d)_{ns} - \sum_{(i \in I_n),b} g_{i,t,b} + \sum_{m \in M_n} \frac{(\theta_{n,t} - \theta_{m,t})}{x_{n,m}} = 0 : \lambda_{n,t}, \ \forall n, \ \forall t \quad (5.26)$$

$$0 \leqslant g_{i,t,b} \leqslant g_{i,b}^{\max} : \mu_{i,t,b}^-, \mu_{i,t,b}^+, \ \forall i, \ \forall t, \ \forall b \quad (5.27)$$

$$0 \leqslant d_{j,t,c} \leqslant d_{j,t,c}^{\max} : \nu_{j,t,c}^-, \nu_{j,t,c}^+, \ \forall j, \ \forall t, \ \forall c \quad (5.28)$$

$$E_t = E_{t-1} + \eta^c s_t^c - s_t^d / \eta^d : \xi_t, \ \forall t \quad (5.29)$$

$$E^{\min} \leqslant E_t \leqslant E^{\max} : \pi_t^-, \pi_t^+, \ \forall t \quad (5.30)$$

$$0 \leqslant s_t^c \leqslant (1 - k_t) s^{\max} : \rho_t^-, \rho_t^+, \ \forall t \quad (5.31)$$

$$0 \leqslant s_t^d \leqslant (1 - k_t) s^{\max} : \sigma_t^-, \sigma_t^+, \ \forall t \quad (5.32)$$

$$E_0 = E_{NT} : \varphi \quad (5.33)$$

$$-F_{n,m}^{\max} \leqslant \frac{\theta_{n,t} - \theta_{m,t}}{x_{n,m}} \leqslant F_{n,m}^{\max} : \beta_{n,m,t}^-, \beta_{n,m,t}^+, \ \forall n, \ \forall m \in M_n, \ \forall t \quad (5.34)$$

$$-\pi \leqslant \theta_{n,t} \leqslant \pi : \gamma_{n,t}^-, \gamma_{n,t}^+, \ \forall n, \ \forall t \quad (5.35)$$

$$\theta_{1,t} = 0 : \delta_t, \ \forall t \quad (5.36)$$

The upper level problem determines the optimal capacity withholding strategy k_t maximizing the profit of the energy storage participant. This problem is subject to the limits of the capacity withholding strategies (5.23) and the lower level problem

(5.24)—(5.35). The latter represents the market clearing process, maximizing the social welfare (5.24), subject to nodal demand-supply balance constraint (5.26) (the Lagrangian multipliers of which constitute the locational marginal prices at each node and each time period), generation and demand limits (5.27)—(5.28), the perceived operational constraints of energy storage (5.29)—(5.33). Limits on transmission line capacities and voltage angles of nodes are enforced through constraints (5.34)—(5.35), respectively. Finally, constraint (5.36) identifies $n=1$ as the reference node.

5.5.2 MPEC Formulation

In order to solve the bi-level model (5.22)—(5.36) of the strategic energy storage participant, the LL problem is replaced by its KKT optimality conditions, which is enabled by the continuity and convexity of the lower level problem. For the conciseness reasons, the full derivation of KKT conditions follows the procedure presented in subsection 4.5.2 and is not repeated here. The resultant MPEC problem is formulated as:

$$\max_{V^{\text{MPEC}}} \sum_t \lambda_{n_s, t}(s_t^d - s_t^c) \tag{5.37}$$

where:

$$V^{\text{MPEC}} = \{k_t, V^{\text{LL-P}}, \lambda_{n,t}, \mu_{i,t,b}^-, \mu_{i,t,b}^+, \nu_{j,t,c}^-, \nu_{j,t,c}^+, \xi_t, \pi_t^-, \pi_t^+, \rho_t^-, \rho_t^+, \sigma_t^-, \sigma_t^+, \varphi,$$

$$\beta_{n,m,t}^-, \beta_{n,m,t}^+, \gamma_{n,t}^-, \gamma_{n,t}^+, \delta_t\} \tag{5.38}$$

subject to:

$$0 \leqslant k_t \leqslant 1, \ \forall t \tag{5.39}$$

$$\lambda_{i,b}^G - \lambda_{(n: i \in I_n), t} - \mu_{i,t,b}^- + \mu_{i,t,b}^+ = 0, \ \forall i, \ \forall t, \ \forall b \tag{5.40}$$

$$-\lambda_{j,t,c}^D + \lambda_{(n: j \in J_n), t} - \nu_{j,t,c}^- + \nu_{j,t,c}^+ = 0, \ \forall j, \ \forall t, \ \forall c \tag{5.41}$$

$$\lambda_{n_s, t} - \rho_t^- + \rho_t^+ - \eta^c \xi_t = 0, \ \forall t \tag{5.42}$$

$$-\lambda_{n_s, t} - \sigma_t^- + \sigma_t^+ + \xi_t / \eta^d = 0, \ \forall t \tag{5.43}$$

$$-\pi_t^- + \pi_t^+ + \xi_t - \xi_{t+1} = 0, \ \forall t < N_T \tag{5.44}$$

$$-\pi_{N_T}^- + \pi_{N_T}^+ + \xi_{N_T} - \varphi = 0 \tag{5.45}$$

$$\sum_{m \in M_n} \frac{\lambda_{n,t} - \lambda_{m,t}}{x_{n,m}} + \sum_{m \in M_n} \frac{\beta_{n,m,t}^+ - \beta_{m,n,t}^+}{x_{n,m}} - \sum_{m \in M_n} \frac{\beta_{n,m,t}^- - \beta_{m,n,t}^-}{x_{n,m}} + \gamma_{n,t}^+ - \gamma_{n,t}^- + (\delta_t)_{n=1} = 0,$$

$$\forall n, \ \forall t \tag{5.46}$$

$$\sum_{(j \in J_n), c} d_{j,t,c} + (s_t^c - s_t^d)_{ns} - \sum_{(i \in I_n), b} g_{i,t,b} + \sum_{m \in M_n} \frac{(\theta_{n,t} - \theta_{m,t})}{x_{n,m}} = 0, \ \forall n, \ \forall t$$
(5.47)

$$E_t = E_{t-1} + \eta^c s_t^c - s_t^d / \eta^d, \ \forall t \tag{5.48}$$

$$E_0 = E_{NT} \tag{5.49}$$

$$\theta_{1,t} = 0, \ \forall t \tag{5.50}$$

$$0 \leq \mu_{i,t,b}^- \perp g_{i,t,b} \geq 0, \ \forall i, \ \forall t, \ \forall b \tag{5.51}$$

$$0 \leq \mu_{i,t,b}^+ \perp (g_{i,b}^{\max} - g_{i,t,b}) \geq 0, \ \forall i, \ \forall t, \ \forall b \tag{5.52}$$

$$0 \leq \nu_{j,t,c}^- \perp d_{j,t,c} \geq 0, \ \forall j, \ \forall t, \ \forall c \tag{5.53}$$

$$0 \leq \nu_{j,t,c}^+ \perp (d_{j,t,c}^{\max} - d_{j,t,c}) \geq 0, \ \forall j, \ \forall t, \ \forall c \tag{5.54}$$

$$0 \leq \pi_t^- \perp (E_t - E^{\min}) \geq 0, \ \forall t \tag{5.55}$$

$$0 \leq \pi_t^+ \perp (E^{\max} - E_t) \geq 0, \ \forall t \tag{5.56}$$

$$0 \leq \rho_t^- \perp s_t^c \geq 0, \ \forall t \tag{5.57}$$

$$0 \leq \rho_t^+ \perp ((1-k_t) s^{\max} - s_t^c) \geq 0, \ \forall t \tag{5.58}$$

$$0 \leq \sigma_t^- \perp s_t^d \geq 0, \ \forall t \tag{5.59}$$

$$0 \leq \sigma_t^+ \perp ((1-k_t) s^{\max} - s_t^d) \geq 0, \ \forall t \tag{5.60}$$

$$0 \leq \beta_{n,m,t}^- \perp \left(F_{n,m}^{\max} + \frac{\theta_{n,t} - \theta_{m,t}}{x_{n,m}} \right) \geq 0, \ \forall n, \ \forall m \in M_n, \ \forall t \tag{5.61}$$

$$0 \leq \beta_{n,m,t}^+ \perp \left(F_{n,m}^{\max} - \frac{\theta_{n,t} - \theta_{m,t}}{x_{n,m}} \right) \geq 0, \ \forall n, \ \forall m \in M_n, \ \forall t \tag{5.62}$$

$$0 \leq \gamma_{n,t}^- \perp (\pi + \theta_{n,t}) \geq 0, \ \forall n, \ \forall t \tag{5.63}$$

$$0 \leq \gamma_{n,t}^+ \perp (\pi - \theta_{n,t}) \geq 0, \ \forall n, \ \forall t \tag{5.64}$$

The objective function of the MPEC is identical to the objective function of the UL problem. The set of decision variables (5.38) includes the decision variables of the UL and the LL problems in (5.25), as well as the Lagrangian multipliers (i.e. the dual variables) associated with the constraints of the LL problem. The KKT optimality conditions of the LL problem correspond to equations (5.40)–(5.64).

5.5.3 MILP Formulation

As mentioned in subsection 4.5.3, the above MPEC formulation is non-linear and thus any solution obtained by commercial solvers is not guaranteed to be globally optimal. In order to address this challenge, the MPEC is transformed to a Mixed-Integer Linear Problem (MILP) which can be efficiently solved using available branch-and-cut solvers. This linearization process is presented in this subsection.

The above MPEC includes two types of non-linearities. The first one involves the bilinear objective function (5.37). In order to linearize it, we propose an approach making use of some KKT conditions. First of all, by making use of the demand-supply balance constraint (5.26) for node n_s, the objective function (5.37) becomes equal to:

$$\max_V \sum_{(j \in J_{n_s}), t, c} \lambda_{n_s, t} d_{j, t, c} - \sum_{(i \in I_{n_s}), t, b} \lambda_{n_s, t} g_{i, t, b} + \sum_{(m \in M_{n_s}), t} \lambda_{n_s, t} \frac{(\theta_{n_s, t} - \theta_{m, t})}{x_{n_s, m}}$$

(5.65)

By multiplying both sides of (5.40) by $g_{i, t, b}$, summing for every $i \in I_{n_s}$, t and b, and rearranging some terms we get:

$$\sum_{(i \in I_{n_s}), t, b} \lambda_{n_s, t} g_{i, t, b} = \sum_{(i \in I_{n_s}), t, b} (\lambda_{i, b}^G g_{i, t, b} - \mu_{i, t, b}^- g_{i, t, b} + \mu_{i, t, b}^+ g_{i, t, b}) \quad (5.66)$$

By multiplying both sides of (5.41) by $d_{j, t, c}$, summing for every $j \in J_{n_s}$, t and c and rearranging some terms we get:

$$\sum_{(j \in J_{n_s}), t, c} \lambda_{n_s, t} d_{j, t, c} = \sum_{(j \in J_{n_s}), t, c} (\lambda_{j, t, c}^D d_{j, t, c} + v_{j, t, c}^- d_{j, t, c} - v_{j, t, c}^+ d_{j, t, c}) \quad (5.67)$$

By multiplying both sides of (5.46) for node n_s by θ_{n_s}, summing for every t and rearranging some terms we get:

$$\sum_{(m \in M_{n_s}), t} \lambda_{n_s, t} \frac{(\theta_{n_s, t} - \theta_{m, t})}{x_{n_s, m}} = - \sum_{(m \in M_{n_s}), t} \beta_{n_s, m, t}^+ \frac{(\theta_{n_s, t} - \theta_{m, t})}{x_{n_s, m}} +$$

$$\sum_{(m \in M_{n_s}), t} \beta_{n_s, m, t}^- \frac{(\theta_{n_s, t} - \theta_{m, t})}{x_{n_s, m}} - \theta_{n_s} \gamma_{n_s, t}^+ + \theta_{n_s} \gamma_{n_s, t}^-$$

(5.68)

By making use of constraint (5.51) and constraint (5.52), equation (5.66) becomes:

$$\sum_{(i \in I_{n_s}), t, b} \lambda_{n_s, t} g_{i, t, b} = \sum_{(i \in I_{n_s}), t, b} (\lambda_{i, b}^G g_{i, t, b} + \mu_{i, t, b}^+ g_{i, b}^{\max}) \quad (5.69)$$

By making use of constraint (5.53) and constraint (5.54), equation (5.67) becomes:

$$\sum_{(j \in J_{n_s}), t, c} \lambda_{n_s, t} d_{j, t, c} = \sum_{(j \in J_{n_s}), t, c} (\lambda_{j, t, c}^D d_{j, t, c} - v_{j, t, c}^+ d_{j, t, c}^{\max}) \quad (5.70)$$

By making use of constraint (5.61)—(5.64), equation (5.68) becomes:

$$\sum_{(m \in M_{n_s}), t} \lambda_{n_s, t} \frac{(\theta_{n_s, t} - \theta_{m, t})}{x_{n_s, m}} = - \sum_{(m \in M_{n_s}), t} (\beta_{n_s, m, t}^+ + \beta_{n_s, m, t}^-) F_{n, m}^{\max} - (\gamma_{n_s, t}^+ + \gamma_{n_s, t}^-) \pi$$

(5.71)

By substituting equations (5.69)—(5.71) into function (5.65), we get the following linear objective function, which replaces the non-linear objective function (5.37):

$$\max_{V} \sum_{(j \in J_{ns}), t, c} (\lambda^{D}_{j,t,c} d_{j,t,c} - v^{+}_{j,t,c} d^{max}_{j,t,c}) - \sum_{(i \in I_{ns}), t, b} (\lambda^{G}_{i,b} g_{i,t,b} + \mu^{+}_{i,t,b} g^{max}_{i,b}) -$$

$$\sum_{(m \in M_{ns}), t} (\beta^{+}_{ns,m,t} + \beta^{-}_{ns,m,t}) F^{max}_{n,m} - (\gamma^{+}_{ns,t} + \gamma^{-}_{ns,t}) \pi \quad (5.72)$$

The second type of non-linearity involves the bilinear terms in the complementarity conditions (5.51)—(5.64). Similarly, as in subsection 4.5.3, the linearization approach proposed in reference [122] is employed to replace these conditions with a set of mixed-integer linear conditions. The set of decision variables of the MILP formulation includes the set (5.38) as well as the auxiliary binary variables introduced for linearizing (5.51)—(5.64).

5.6 Case Studies

5.6.1 Test Data and Implementation

The examined studies demonstrate the ability of energy storage to exercise market power through capacity withholding in the test market with day-ahead horizon and hourly resolution, operating over a 16-node model of the GB transmission network (Figure 4.12). As mentioned in subsection 4.6.1, the transmission network is divided into two regions, with nodes 1-6 correspond to Scotland while nodes 7-16 correspond to England, and interconnected by transmission link (6,7).

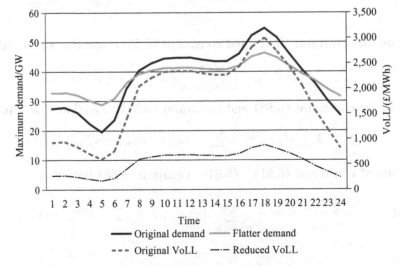

Figure 5.6 Profiles of maximum system demand and VoLL

The market includes 7 generation participants, with their linear/quadratic cost coefficients, maximum output limits[127] and locations given in Table 4.1. The market also includes 13 demand participants, with the location and relative size (expressed as % of the total system demand and assuming that it remains identical for every time period) presented in Table 4.2. As discussed in subsection 5.3.2, the coefficients of the demand side's benefit function and maximum baseline demand are time-differentiated parameters, following the daily pattern of consumers' activities. Two alternative cases for the profile of the maximum demand in the whole system are considered (Figure 5.6): the "Original demand" profile is based on a typical winter day demand profile of the GB system (Figure 4.13), while the "Flatter demand" profile exhibits the same shape and daily energy consumption as the "Original demand" profile, but exhibits a smaller difference between peak and off-peak levels. The hourly values of the linear benefit coefficient (i.e. VoLL) are assumed identical for each node. Two alternative cases for these values are considered (Figure 5.6): the "Original VoLL" case is based on values used in the GB system (Figure 4.13), while the "Reduced VoLL" case represents a scenario of higher price responsiveness of the demand side.

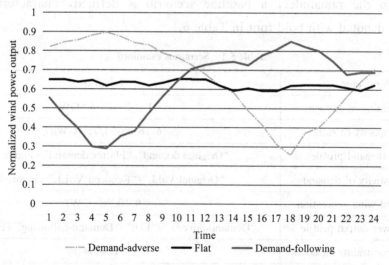

Figure 5.7 Normalized wind power output profiles

Driven by the recent efforts to decarbonize electricity generation, the market also includes wind generators (assumed connected to node 5 in Scotland), which constitute the dominant renewable energy technology in GB. The operating costs of these generators are assumed zero, implying that they are prioritized in the merit order and the 7 generation participants of Table 4.1 need to satisfy the remaining or net demand (i.

e. difference between demand and wind power output). Three alternative cases for the profile of the normalized wind power output (i.e. power output divided by installed capacity) in the system are considered, illustrated in Figure 5.7. The "Demand-following" profile follows the daily pattern characterizing the system demand (Figure 5.6), implying high wind generation during peak demand periods and low wind generation during off-peak demand periods. The "Demand-adverse" profile exhibits an opposite pattern, with high wind generation during off-peak demand periods and low wind generation during peak demand periods. The "Flat" profile represents a situation where wind generation remains relatively constant throughout the day. For comparison purposes, the daily wind energy production is the same for all three profiles.

In order to comprehensively analyse the impact of different factors on the ability of energy storage to exercise market power, different scenarios have been examined regarding i) the power rating and energy capacity of energy storage, ii) the daily profile and price elasticity of the demand side, iii) the installed capacity and daily output profile of wind generation, and iv) the location of energy storage in the network, in case the network congestion occurs. These scenarios are presented in Table 5.1. For ease of reference in the remainder, a baseline scenario is defined, characterized by the parameters denoted with bold font in Table 5.1.

Table 5.1 Scenarios examined

Parameter	Scenarios
Power rating of storage	1/2/3/4/5/**6** (GW)
Energy capacity of storage	**6**/18/36/54/72 (GWh)
System demand profile	**"Original demand"**/"Flatter demand" (Figure 5.6)
Price elasticity of demand	**"Original VoLL"**/"Reduced VoLL" (Figure 5.6)
Capacity of wind generation	**0**/10/20 (GW)
Wind power output profile	"Demand-adverse"/"Flat"/"Demand-following" (Figure 5.7)
Network constraints included	Yes/**No**
Location of storage	Node 5 in Scotland/Node 16 in England

Apart from the scenario-dependent values of the power rating and energy capacity, the assumed values for the remaining operating parameters of energy storage are the same in every scenario and are presented in Table 4.3.

As discussed in subsection 5.3.1 and subsection 5.3.2, the quadratic generation cost and demand benefit functions are approximated by piecewise linear functions, consisting of

20 blocks each. The MILP formulation was coded and solved using the optimization software FICO™ Xpress[127] on a computer with a 6-core 3.47 GHz Intel(R) Xeon(R) X5690 processor and 192 GB of RAM. The average computational time required for solving the MILP across all examined scenarios was approximately 30 s.

5.6.2 Quantifying the Optimal Extent of Capacity Withholding by Energy Storage

Considering the baseline scenario, three cases concerning the behaviour of energy storage in the market are compared: i) competitive behaviour, where the capacity withholding strategies are set equal to $k_t = 0$, $\forall t$, ii) optimized strategic behaviour, where the capacity withholding strategies are optimized through the proposed model, and iii) naïve strategic behaviour, where the capacity withholding strategies are set equal to a high value (in particular $k_t = 0.8$, $\forall t$).

Figure 5.8, Figure 5.9 and Table 5.2 present the hourly storage charging/discharging schedule (indicated by negative and positive values respectively), the hourly market clearing prices, and the daily storage profit respectively, under each of the three aforementioned cases.

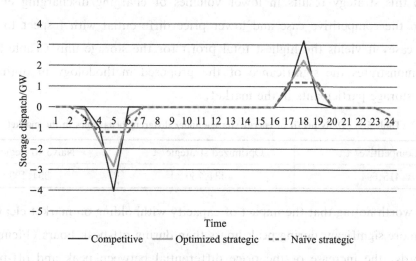

Figure 5.8 Hourly storage charging/discharging schedule under different energy storage behaviours in the market

Under competitive behaviour, the energy charged and discharged by the storage unit during off-peak and peak hours respectively is the highest (leading to a flatter net demand profile), but the price differential between peak and off-peak hours is the lowest. Under naïve strategic behaviour, the energy charged and discharged by the

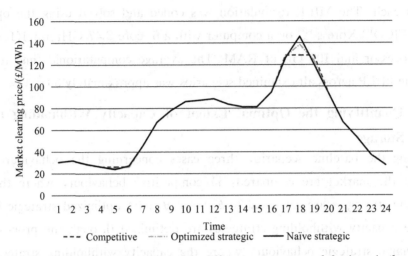

Figure 5.9 Hourly market clearing prices under different energy storage behaviours in the market

storage unit is the lowest, but the price differential between peak and off-peak hours is the highest.

Under optimized strategic behaviour, the storage unit determines its capacity withholding strategies based on the optimal trade-off between these two effects. Although this strategy results in lower volumes of charging/discharging energy with respect to the competitive case and lower price differential with respect to the naïve strategic case, it yields the highest total profit for the storage unit (Table 5.2). This result demonstrates the significance of the proposed methodology in optimizing the profit of storage participants in the market.

Table 5.2 Daily profit of energy storage under different behaviours in the market

Competitive/ £	Optimized strategic/ £	Naïve strategic/ £
415,694	434,760	354,103

It is also worth noting that the impact of capacity withholding on market clearing prices is much more significant during peak hours than during off-peak hours (Figure 5.9). In other words, the increase of the price differential between peak and off-peak hours driven by capacity withholding is mainly associated with the larger increase of prices during peak hours rather than the smaller reduction of prices during off-peak hours (as also qualitatively illustrated in Figure 5.3). This effect emerges due to the fact that the slope of the generation offering curve exhibits steeper slopes at higher demand levels (Figure 5.3 and Table 4.1) and explains some of the trends observed in the following sections.

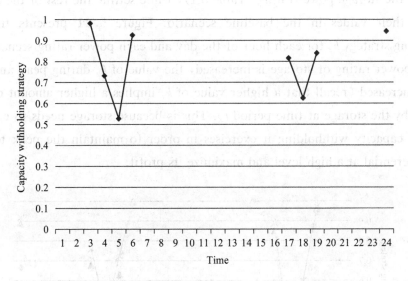

Figure 5.10 Capacity withholding strategies under optimized strategic behaviour

Figure 5.10 presents the withholding strategy k_t under optimized strategic behaviour. It should be noted that during hours when storage is neither charged nor discharged (mid-peak hours in general, e.g. hours 1–2, 7–16, and 20–23 in Figure 5.8), the value of k_t does not affect the storage profit, as verified by additional tests conducted by the authors; thus, Figure 5.10 and similar figures in the remainder of this chapter do not present any value for k_t in these hours. The value of k_t affects storage profit during (off-peak) hours when storage is charged (e.g. hours 3–6 in Figure 5.8) and (peak) hours when storage is discharged (e.g. hours 17–19 in Figure 5.8).

The optimal extent of exercised capacity withholding is not the same across all off-peak and peak hours. It is noticed that the hours exhibiting the lowest extent of capacity withholding in the off-peak and the peak time window are hours 5 and 18 respectively. The reason behind this trend is associated with the negative effect of capacity withholding on the volume of energy sold by the storage unit in the market, discussed in subsection 5.4.2. Since the storage unit charges and discharges most of its energy at hours 5 and 18 respectively (Figure 5.8), it is motivated to act more truthfully during these hours.

5.6.3 Impact of Storage Size

Two case studies are investigated in this subsection. In the first one, the optimal decision-making problem of strategic storage has been solved for different scenarios

147

regarding the storage power rating (Table 5.1), while setting the rest of the parameters equal to their values in the baseline scenario. Figure 5.11 presents the optimal withholding strategy k_t for each hour of the day and each power rating scenario. As the (actual) power rating of storage is increased, the value of k_t during peak and off-peak hours is increased (recall that a higher value of k_t implies a higher amount of capacity withheld by the storage at time period t). This is because storage needs to enhance the extent of capacity withholding it exercises in order to maintain the peak to off-peak price differential at a high level and maximize its profit.

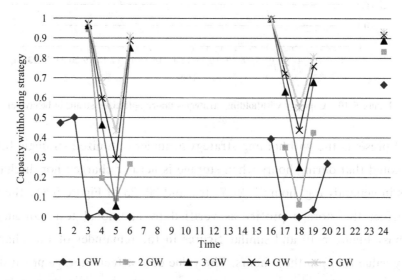

Figure 5.11 Impact of power rating on optimal capacity withholding strategies of energy storage

As discussed in subsection 5.2.2 and subsection 5.4.2, the exertion of market power by energy storage increases its profit and decreases social welfare, compared to the case it behaves competitively. Therefore, two cases are compared: i) a case of perfectly competitive market (indicated by the superscript c in the remainder), where the energy storage behaves competitively at all time periods, i.e. $k_t = 0$, $\forall t$, and ii) a case of imperfect market (indicated by the superscript s in the remainder), where the capacity withholding strategies of the storage are determined based on the developed game-theoretic model (Section 5.5). Relevant market power metrics (subsection 4.6.2) are employed in this and the following subsections in order to quantitatively characterise the extent of market power exercised by the energy storage. Figure 5.12 presents the impact of storage power rating on the *storage profit deviation index* (SPDI) (5.73) and the *market inefficiency index* (MII) (5.74) driven by the exercise of capacity withholding.

$$SPDI = \frac{\sum_t SP_t^s - \sum_t SP_t^c}{\sum_t SP_t^c} \times 100\% \quad (5.73)$$

$$MII = \frac{\sum_t SW_t^s - \sum_t SW_t^c}{\sum_t SW_t^c} \times 100\% \quad (5.74)$$

As the power rating increases from 1 GW to 4 GW, the SPDI and the absolute values of MII are increased, because its larger size enables storage to exercise more market power. Under competitive storage behaviour, increase of the storage power rating beyond 4 GW cannot further flatten the demand profile and therefore does not change further the schedule of storage. As a result, the power rating revealed by the storage unit to the market under strategic behaviour also remains unchanged; this is justified by the fact that the revealed rating, i.e. $(1-k_t) s^{\max}$, is identical for $s^{\max} = 4$ GW and $s^{\max} = 5$ GW at every hour, as illustrated in Figure 5.11. Therefore, the SPDI and the absolute values of MII driven by the exercise of market power exhibit a saturation effect for a power rating higher than 4 GW (Figure 5.12).

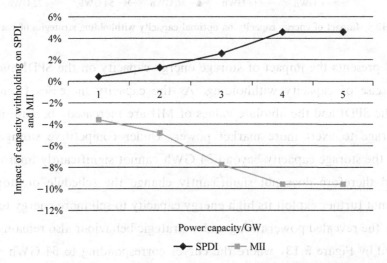

Figure 5.12 Impact of power rating on storage profit deviation index (SPDI) and market inefficiency index (MII) driven by the exercise of capacity withholding

In the second case study, the optimal decision-making problem of strategic storage has been solved for different scenarios regarding the storage energy capacity (Table 5.1), while setting the rest of the parameters equal to their values in the baseline scenario. Figure 5.13 presents the optimal withholding strategy k_t for each hour of the day and each energy capacity scenario. In contrast to the impact of increased power rating examined before, as the capacity increases, the value of k_t is decreased. The reason behind this trend is associated with the negative effect of capacity withholding on the volume of energy sold by the storage unit in the market. As its capacity increases, storage is motivated to act more truthfully in order to exploit its higher energy content and sell more energy in the market.

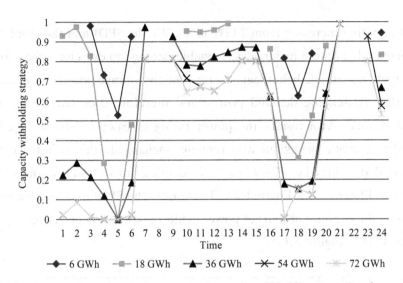

Figure 5.13 Impact of energy capacity on optimal capacity withholding strategies of energy storage

Figure 5.14 presents the impact of storage energy capacity on the SPDI and MII driven by the exercise of capacity withholding. As the capacity increases from 6 GWh to 54 GWh, the SPDI and the absolute values of MII are increased, because its larger size enables storage to exert more market power. Under competitive storage behaviour, increase of the storage capacity beyond 54 GWh cannot significantly improve the social welfare and therefore does not significantly change the schedule of storage, as the storage cannot further exploit its high energy capacity to sell more energy to the market. As a result, the revealed power rating under strategic behaviour also remains unchanged, as illustrated by Figure 5.13, where the curves corresponding to 54 GWh and 72 GWh almost coincide. Therefore, the SPDI and the absolute values of MII driven by the exercise of market power exhibit a saturation effect for a capacity higher than 54 GWh (Figure 5.14).

5.6.4 Impact of the Characteristics of the Demand Side

Two case studies are examined in this subsection. In the first one, the optimal decision-making problem of strategic storage has been solved for the two different demand

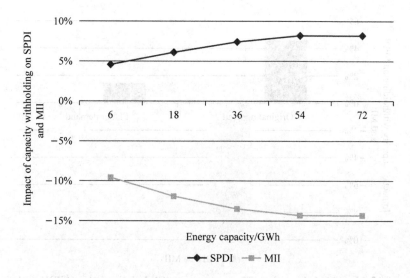

Figure 5.14 Impact of energy capacity on storage profit deviation index (SPDI) and market inefficiency index (MII) driven by the exercise of capacity withholding

profiles of Figure 5.6 ("Original demand" and "Flatter demand" profile) while setting the rest of the parameters equal to their values in the baseline scenario. Since the impact of capacity withholding on the price differential between peak and off-peak hours is mainly associated with the increase of prices during peak hours (Figure 5.3), this impact is higher under the "Original demand" profile which exhibits a higher peak demand. As a result, the SPDI and the absolute values of MII are both higher under the "Original demand" than under the "Flatter demand" profile (Figure 5.15). This result indicates that demand response measures flattening the demand profile by rescheduling consumers' demand from peak towards off-peak periods will have a positive effect in mitigating the extent of market power exercised by strategic storage.

In the second case study, the optimal decision-making problem of strategic storage has been solved for the two different VoLL cases of Figure 5.6 ("Original VoLL" and "Reduced VoLL") while setting the rest of the parameters equal to their values in the baseline scenario. It should be noted that, for the same demand profile, a lower VoLL implies a scenario of higher consumers' price elasticity, that is, a lower level of electricity will be consumed at any market clearing prices.

Figure 5.16 presents the impact of the VoLL on the SPDI and MII driven by the exercise of capacity withholding. A higher price elasticity (lower VoLL) leads to a reduction across SPDI and the absolute values of MII, due to the combination of two effects: i) the peak demand is reduced and therefore capacity withholding has a lower impact on

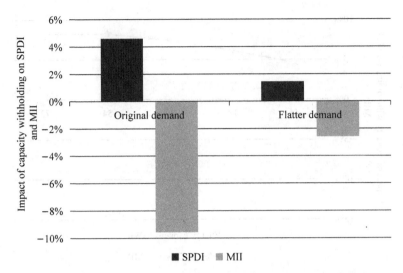

Figure 5.15 Impact of demand profile on storage profit deviation index (SPDI) and market inefficiency index (MII) driven by the exercise of capacity withholding

the price differential between peak and off-peak hours, and ii) consumers' demand becomes more sensitive to price changes and therefore capacity withholding yields a higher reduction to the volume of energy sold by the storage unit. This result indicates that demand response measures enhancing the consumers' price responsiveness will have a positive effect in limiting the extent of market power exercised by strategic storage.

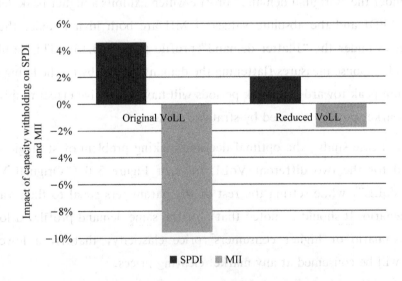

Figure 5.16 Impact of demand's price elasticity on storage profit deviation index (SPDI) and market inefficiency index (MII) driven by the exercise of capacity withholding

5.6.5 Impact of the Characteristics of Wind Generation

Two case studies are examined in this subsection. In the first one, the optimal decision-making problem of strategic storage has been solved for different scenarios regarding the installed wind generation capacity (Table 5.1), while assuming that the wind power output follows the "Flat" profile of Figure 5.7 and setting the rest of the parameters equal to their values in the baseline scenario. As the wind generation capacity increases, the (net) peak demand is reduced and therefore capacity withholding has a lower impact on the price differential between peak and off-peak hours. As a result, the SPDI and the absolute values of MII are reduced as the wind generation capacity increases (Figure 5.17).

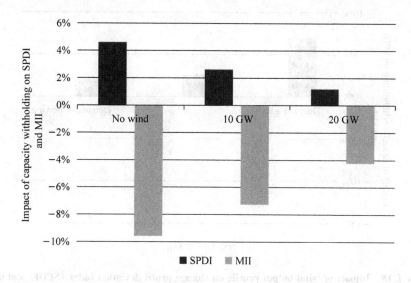

Figure 5.17   Impact of wind generation capacity on storage profit deviation index (SPDI) and market inefficiency index (MII) driven by the exercise of capacity withholding

In the second case study, the optimal decision-making problem of strategic storage has been solved for the three different wind power output profiles of Figure 5.7 ("Demand-following", "Flat", and "Demand-adverse" profiles), while assuming an installed wind generation capacity of 10 GW and setting the rest of the parameters equal to their values in the baseline scenario. Similar trends with the ones observed in subsection 5.6.4 regarding the impact of the shape of the demand profile are observed here. The "Demand-following" wind output profile has similar effects as the "Flatter demand" profile, since it leads to a flatter (net) demand profile. The "Demand-adverse" wind output profile has similar effects as the "Original demand" profile, since it leads to a (net) demand profile characterized by a higher difference between peak and off-peak levels. The "Flat" wind output profile constitutes an intermediate case. As a result, "Demand-adverse" case is characterized by the highest SPDI and the absolute values of MII driven by the exercise of market power, followed by the "Flat" and "Demand-following" cases (Figure 5.18). This result demonstrates that the extent of market power exercised by strategic storage also significantly depends on the output profile of wind generation. Therefore, in future systems with high wind penetration levels, storage operators will require suitable forecasting and stochastic optimization tools in order to devise their strategies in the market fully taking into account the inherent variability and uncertainty associated with wind production.

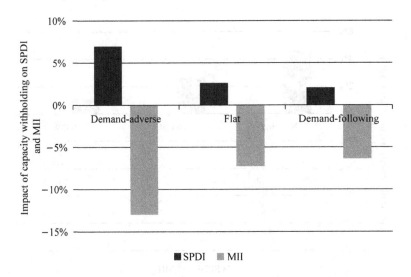

Figure 5.18 Impact of wind output profile on storage profit deviation index (SPDI) and market inefficiency index (MII) driven by the exercise of capacity withholding

5.6.6 Impact of Storage Location

In this subsection, network constraints are taken into account and the optimal decision-making problem of strategic storage has been solved for the following four cases:

U-ES-SC: The network is uncongested and an ES unit (whose operational parameters set equal to their values in the baseline scenario) is connected to node 5 in Scotland.

U-ES-EN: The network is uncongested and the same ES unit is connected to node 16 in England.

C-ES-SC: Line (6, 7) is congested and the same ES unit is connected to node 5 in

Scotland.

C-ES-EN: Line (6, 7) is congested and the same ES unit is connected to node 16 in England.

When the network is uncongested, the market outcome is identical with the case where network constraints are neglected, irrespectively of the storage location. Consequently, case U-ES-SC and case U-ES-EN exhibit the same values of indexes (Figure 5.19). When the network is congested though, the location of storage affects the market outcome significantly, and consequently affects its ability to exercise market power (Figure 5.19). The reason behind this result is associated with the fact that network congestion decouples to some extent the market conditions at different locations. In the examined setting, Scotland is characterized by cheaper generation and higher demand (since it includes largest proportion of base units and 14% of the system demand) than England (which includes largest proportion of mid and peak units and 86% of the system demand). As a result, during periods when the line is congested, England exhibits higher prices than the ones observed in the uncongested case, while Scotland exhibits lower prices than the ones observed in the uncongested case. Therefore, the impact of storage capacity withholding on off-peak to peak price differential, and the resulting SPDI and the absolute values of MII are significantly higher when storage is located in England and significantly lower when it is located in Scotland (Figure 5.19).

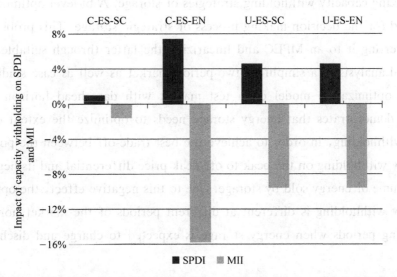

Figure 5.19 Impact of storage location and network congestion on storage profit deviation index (SPDI) and market inefficiency index (MII) driven by the exercise of capacity withholding

5.7 Conclusions

Previous work analysing market opportunity of demand side has demonstrated the ability of the latter to strategically increase their surplus by submitting bids lower than their actual marginal benefit, however the option of which to exercise market power by revealing less flexibility potential to the market is not addressed. Furthermore, although previous studies have demonstrated the ability of large energy storage units to exercise market power by withholding their capacity, some have adopted modelling approaches exhibiting certain limitations and have not properly analysed the dependency of the extent of exercised market power on ES and system parameters.

This chapter aims at addressing the above challenges. Regarding the demand side, qualitative analysis through a price-quantity graph in a simplified two-period market has demonstrated that strategically revealing less time-shifting flexibility has a more dominant negative impact on demand payment during the peak period than its positive impact during the off-peak period, which leads to an overall higher demand payment. As a result, consumers will find it financially unattractive to behave in such a strategic fashion.

Concerning energy storage, this chapter develops a multi-period game-theoretic model for optimizing capacity withholding strategies of storage. A bi-level optimization model is proposed for the decision-making process of strategic storage. This problem is solved after converting it to an MPEC and linearizing the latter through suitable techniques. Theoretical analysis in a simplified two-period market as well as case studies with the developed optimization model on a test market with day-ahead horizon and hourly resolution demonstrates that energy storage needs to optimize the extent of exercised capacity withholding, in order to achieve the best trade-off between the positive effect of capacity withholding on the peak to off-peak price differential and its negative effect on the volume of energy sold by storage. Due to this negative effect, the optimal extent of capacity withholding is different at different periods of the market horizon, being lower during periods when energy storage is expected to charge and discharge higher energy.

Case studies have adopted relevant indexes to measure the extent of market power exercised by the energy storage. The impact of the size of energy storage (in terms of its power rating and energy capacity) on the extent of exercised capacity withholding and the resulting storage profit increase and social welfare loss is analyzed. A higher power rating increases the extent of exercised capacity withholding as energy storage attempts to maintain the peak to off-peak price differential at a high level. A higher energy capacity reduces the extent of capacity withholding as energy storage is motivated to act more truthfully in order to exploit its higher energy content and sell more energy in the market. Nevertheless, both a higher power rating and a higher energy capacity increase the additional profit made by energy storage and the social welfare loss (with respect to the case where storage acts competitively), since they increase the ability of energy storage to affect the market prices. This trend however exhibits a saturation effect, as increase of the power rating and the energy capacity beyond a certain level does not affect the market outcome under neither competitive nor strategic storage behaviour.

Furthermore, case studies have analysed the impact of the generation and demand characteristics of the market on the extent of market power exercised by energy storage. Concerning the demand side, a higher price elasticity and a flatter demand profile reduce the additional profit made by energy storage and the social welfare loss. This result indicates that the envisaged realization of demand response measures enhancing the consumers' price responsiveness and flattening the demand profile by rescheduling consumers' demand towards (lower-priced) off-peak periods will have a positive effect in mitigating the extent of market power exercised by strategic storage. Regarding the generation side, a higher penetration of wind generation suppresses price levels and therefore generally limits the extent of market power exercised by strategic storage. Finally, the case studies have demonstrated that the location of storage affects its ability to exercise market power if the network over which the market operates is congested. In this case, the market power potential of storage is significantly increased when it is located in areas with more expensive generation and higher demand.

Chapter 6
Conclusions and Future Work

6.1 Conclusions

The emerging Smart Grid paradigm has paved the way for the wide introduction of flexible demand (FD) and energy storage (ES) technologies in power systems, with significant economic, technical, and environmental benefits that will facilitate efficient transition to the low-carbon future. In the deregulated energy sector, the realization of the significant FD and ES flexibility potential should be coupled with their suitable integration in electricity markets. Previous studies have proposed market clearing mechanisms considering FD and ES participation and demonstrated their impacts on the system operation. However, these studies have neglected fundamental market complexities, such as modelling and pricing FD non-convexities as well as modelling and analysing the role of FD and ES in imperfect markets.

Firstly, previous work on market with non-convexities has identified non-convexities associated with the generation side of electricity markets, and proposed different approaches to address the resultant inconsistency and surplus sub-optimality effects. Despite the recent research and industry interest in FD technologies due to their significant economic and environmental potential, the demand side has not been examined from the same perspective. Strong motives therefore arise for systematically identification of these FD non-convexities and demonstration of their relation with potential FD inconsistency and surplus sub-optimality effects. It also necessitates appropriate modification in previously proposed pricing mechanisms for addressing non-convexities of the generation side to account for FD non-convexities.

Secondly, the role of the demand side in imperfect electricity markets has been previously investigated in terms of the effect of its own-price elasticity on electricity producers' ability to exercise market power. However, the concept of own-price elasticity cannot fully capture consumers' flexibility regarding energy use, as the latter mainly involves shifting of loads' operation in time. This time-shifting flexibility will be enhanced in the future with the increasing penetration of various FD and ES technologies, envisaged by the Smart Grid paradigm. Although numerous recent studies have investigated the impact of this flexibility on assorted aspects of power system operation and investment, its role in oligopolistic electricity markets has not been explored yet.

Furthermore, the ability of FD and ES to exercise market power needs to be better understood. Regarding the former, previous work has demonstrated that large consumers can strategically increase their surplus by submitting bids lower than their actual marginal benefit, while other market power potential of FDs are not explored. Concerning the latter, although previous studies have demonstrated the ability of large ES units to exercise market power by withholding their capacity, some have adopted modelling approaches exhibiting certain limitations and have not properly analysed the dependency of the extent of exercised market power on ES and system parameters.

This monograph aims at addressing the aforementioned challenges through original contributions associated with the development, analysis and testing of suitable models, methods, and examples to deal with the challenges associated with modelling and pricing FD non-convexities as well as modelling and analysing the role of FD and ES in imperfect markets.

Modelling and pricing FD non-convexities

Detailed operational models are derived to represent two different types of FDs, capturing the largest part of flexible load characteristics in the related literature: *continuously-controllable* FD (CCFD) and *fixed-cycle* FD (FCFD). The power demand of a CCFD can be continuously adjusted between a minimum and a maximum limit when the FD is active. FCFDs involve operating cycles which comprise a sequence of phases occurring at a fixed order, with fixed duration and fixed power consumption, which cannot be altered.

Two different demand flexibility potential are modelled. The first is associated with the ability to completely forgo demand activities and the second is associated with the ability to redistribute the total electrical energy requirements of activities across time. All FDs are assumed to exhibit both flexibility potentials.

Based on a thorough examination of the characteristics of the FD operational models, non-convexities of FD are identified, including options to forgo demand activities, minimum power levels associated with CCFDs as well as discrete power levels associated with FCFDs. The FD non-convexities are different to the respective non-convexities associated with the generators given their significant operational differences.

The relation of these FD non-convexities with schedule inconsistency and surplus sub-optimality effects is demonstrated through simple one- and two-time period examples and a larger case study with day-ahead horizon and hourly resolution. The analysis reveals that in the presence of FD non-convexities and under the traditional marginal pricing mechanism, the FDs concentrate at the lowest-priced hours within their scheduling period under self-scheduling, which is not consistent with the centralized schedule involving an as-flat-as-possible total demand profile. The above inconsistencies are then translated into FD surplus losses.

Generalized uplift and convex hull pricing approaches addressing these schedule inconsistency and surplus sub-optimality effects are suitably extended to account for FD non-convexities. Concerning the former, generalized uplift functions for FD participants are proposed, which include FD-specific terms and constitute additional benefits or payments for the FDs. The structure of the generalized uplift function of FDs is not the same with the respective function for generators, due to their different non-convexities and resulting surplus sub-optimality effects. The parameters of these functions along with the electricity prices are adjusted to achieve consistency for every FD participant. A new rule is introduced for equitable distribution of the total generators' profit loss and FDs' utility loss compensation among market participants. Regarding the latter, convex hull prices are calculated as the Lagrangian multipliers optimizing the dual problem of the market clearing problem considering FD participation. It is demonstrated that convex hull prices are flattened at periods when FD is scheduled to eliminate surplus sub-optimality associated with the FD ability to redistribute energy requirements across time.

Modelling and analysing the role of FD and ES in imperfect markets

The first contribution of this research lies in providing for the first time both theoretical and quantitative evidence of the beneficial impact of FD and ES in limiting market power exercised by strategic electricity producers.

Theoretical explanation of this impact is presented through a price-quantity graph on a simplified two-period market. It is demonstrated that the operation of FD and ES drives a demand profile flattening effect by reducing the peak demand while increasing the off-peak demand. This in turn reduces the price increment (driven by the exercise of market power of strategic producers) at the peak period while increases it at the off-peak period. However, the price increment reduction at the peak period is more prominent than its increase at the off-peak period, due to the larger slope of the marginal cost curve of generation. This effect implies that the deployment of FD and ES results in an overall reduction of strategic producers' ability of to manipulate market prices.

Quantitative analysis is supported by a multi-period *equilibrium programming* model of the oligopolistic market setting. The decision-making process of each strategic producer is modelled through a bi-level optimization problem. The upper level represents the profit maximization problem of the producer and the lower level represents endogenously the market clearing process, accounting for the time-coupling operational constraints of FD and ES, and network constraints. This bi-level problem is converted to a *Mathematical Program with Equilibrium Constraints* (MPEC), by replacing the lower level problem by its equivalent *Karush-Kuhn-Tucker* (KKT) optimality conditions. By employing suitable linearization techniques, this non-linear MPEC is in turn converted to a *Mixed-Integer Linear Program* (MILP) which can be solved using available commercial solvers. An *iterative diagonalization* approach is employed in order to determine market equilibria resulting from the interaction of multiple independent strategic producers.

Case studies with the developed model on a test market reflecting the general generation and demand characteristics of the GB system have quantitatively demonstrated the benefits of FD and ES in limiting producers' market power, by employing relevant indexes from the literature. In cases without network congestion, the location of FD and ES flexibility does not have an impact on producers' market power exercise, but increasing FD flexibility and size of ES is shown to i) reduce strategic producers' ability to manipulate market prices, ii) reduce strategic producers' additional profit driven by the exercise of market power, iii) allow consumers to more efficiently preserve their economic surplus against producers' strategic behavior, and iv) reduce the social welfare loss and thus enhance the overall efficiency of the market. In cases with network congestion, FD and ES flexibility still has an overall positive impact on market efficiency, but the extent of this benefit as well as the impact on producers and demands at different areas depends on the location of FD and ES in the network.

The second contribution of this research lies in exploring the market power potential of FD and ES.

Regarding FD, its ability to exercise market power by revealing less time-shifting flexibility to the market is qualitatively analysed through a price-quantity graph in a simplified two-period market. It is shown that the competitive demand shifting behaviour enables a net demand profile flattening effect by reducing/increasing demand at the peak/off-peak period respectively. By acting strategically and reporting less time-shifting flexibility to the market, the demand participant limits this flattening effect on the system demand profile, since it increases/reduces less the demand at the off-peak/peak period respectively. This strategic action has a more dominant negative impact on demand payment at the peak period (due to the larger slope of the marginal cost curve of generation) than its positive impact at the off-peak period, resulting in an overall higher demand payment. As a result, consumers will find it financially unattractive to behave in such a strategic fashion.

Concerning ES, this monograph develops a multi-period game-theoretic model for optimizing capacity withholding strategies of ES. The decision-making process of strategic ES is modelled through a bi-level optimization problem. The upper level represents the profit maximization objective of the strategic ES and the lower level represents endogenously the market clearing process, accounting for the network constraints. This bi-level problem is converted to a MPEC, by replacing the lower level problem by its equivalent KKT optimality conditions. Suitable linearization technique is proposed to convert this non-linear MPEC to a MILP which can be efficiently solved using available commercial solvers.

Theoretical analysis in a simplified two-period market as well as case studies with the developed optimization model on a test market with day-ahead horizon and hourly resolution demonstrates that ES needs to optimize the extent of exercised capacity withholding, in order to achieve the best trade-off between the positive effect of capacity withholding on the peak to off-peak price differential and its negative effect on the volume of energy sold by ES. Due to this negative effect, the optimal extent of capacity withholding is different at different periods of the market horizon, being lower during periods when ES charges and discharges higher energy.

Case studies have adopted relevant indexes to measure the extent of market power exercised by the energy storage. The impact of the size of energy storage (in terms of its power rating and energy capacity) on the extent of exercised capacity withholding and the resulting storage profit increase and social welfare loss is analyzed. A higher power rating increases the extent of exercised capacity withholding as ES attempts to maintain the peak to off-peak price differential at a high level. A higher energy capacity reduces the extent of capacity withholding as ES is motivated to act more truthfully in order to exploit its higher energy content and sell more energy. Nevertheless, both a higher power rating and a higher energy capacity increase the additional profit made by ES and the social welfare loss (with respect to the case where ES acts competitively), since they increase the ability of ES to affect market prices. This trend however exhibits a saturation effect, as increase of the power rating and the energy capacity beyond a certain level does not affect the market outcome under neither competitive nor strategic ES behaviour.

Case studies also analyse the impact of the generation and demand characteristics of the market on the extent of market power exercised by ES. Concerning the demand side, a higher price elasticity and a flatter demand profile reduce the additional profit made by ES and the social welfare loss. This result indicates that the envisaged realization of demand response measures enhancing the consumers' price responsiveness and flattening the demand profile by rescheduling consumers' demand towards off-peak periods will have a positive effect in limiting the extent of market power exercised by ES. Regarding the generation side, a higher penetration of wind generation suppresses price levels and therefore generally mitigates the extent of market power exercised by strategic ES.

Finally, the studies demonstrate that the location of ES affects its ability to exercise market power if the network over which the market operates is congested. In this case, the market power potential of ES is significantly increased when it is located in areas with more expensive generation and higher demand.

6.2 Future work

6.2.1 Modelling and Pricing Flexible Demand Non-Convexities

Future work regarding this aspect of research should include the following:

- Comprehensively investigate the strengths and weaknesses of generalized uplift and convex hull pricing methodologies, considering both generation and FD market participation in different scenarios for the generation system composition and the penetration and characteristics of flexible demand technologies in the future. For example, suitable case studies should be carried out to examine how the surpluses of generation and FD technologies are affected by their flexibility extent (e.g. minimum stable generation and ramp rate for generators, storage capacity for FD, etc.) under the proposed pricing mechanisms.

- As comprehensively discussed in Chapter 3, the convex hull pricing approach retains uniform pricing, which is characterized by simplicity and transparency. However, it necessitates that the entire compensation of participants' surplus losses is entirely charged to the inflexible demand. This clearly implies that inflexible demand participants are not treated equitably, which is the main weakness of this approach. The generalized uplift approach yields a more equitable distribution of the total surplus loss compensation among the market participants. However, it involves participant-specific pricing terms that may be considered discriminating and non-transparent in real markets. This discriminating, non-transparent nature is the main weakness of the generalized uplift approach. In this context, the potential of combining the transparency and minimum surplus loss property of convex hull pricing with the equitable distribution property of generalized uplifts through a hybrid approach should be explored.

- The introduction of the above pricing/uplift mechanisms can spawn gaming strategies by the independent market players. The potential risks associated with gaming under these mechanisms, as well as remedies for mitigating excessive gaming are among the subjects of future research.

- Extend these mechanisms to account for the network effect, and produce appropriate locational incentives in form of prices/uplift payments that coordinate individual surplus-driven participants to the centralized system solution based on centralized optimal power flow.

- The concept of "fairness" involves the allocation of total surplus loss among different participants and how their surpluses are altered with respect to the centralized solution. However, this concept becomes ambiguous to define and justify when the network effects are accounted for, as it may be unfair for participants located at different nodes of the network to contribute to compensate each other's surplus loss. Future work should extend the proposed equitable surplus loss sharing rule to ensure fairness in the network over which the market operates.

- Include reserve requirements in the scheduling problem and examine reserve pricing mechanisms. This task will first include the modelling of reserve provision not only by generators but also by the FD side. Then, different pricing mechanisms — namely energy only against energy and reserve pricing — will be compared with respect to the profits of different generation and FD participants and different levels of wind in the system. However, one of the fundamental characteristics of a reserve market is the uncertainty of reserve exercised by the contracted participants, depending on the real-time system conditions. In order to capture this uncertainty, suitable probabilistic/stochastic modelling approaches need to be incorporated in the proposed scheduling and pricing mechanisms.

6.2.2 Modelling and Analysing the Role of Flexible Demand and Energy Storage in Imperfect Markets

- Future work aims at enhancing the presented model of Chapter 4 in two directions. Firstly, the generic, technology-agnostic of demand shifting flexibility employed in Chapter 4 will be replaced by more detailed models accounting for the technology-specific operational constraints of different residential and commercial flexible demand technologies, including electric vehicles, electric heat pumps and smart appliances. This will enable a comprehensive analysis of the value of different demand response initiatives in imperfect electricity markets. In addition, different forms of demand side's benefit function will be investigated. For example, benefit functions can model a case where demand shifting is costly to the consumers. In this case, consumers would need to optimize the amount of load shifting accounting for the associated shifting costs. This in turn may affect the value of demand shifting in imperfect markets.

- Secondly, the developed equilibrium programming model as well as similar models in the existing reference [47]—[55], [117]—[118] neglect the complex unit commitment constraints of the generation side, due to their inherent inability to deal with binary decision variables in the lower level of the producers' bi-level optimization problems. However, these complex operating properties may affect the market outcome and the value of demand flexibility, as the latter may have a significant impact on the scheduling pattern of flexible generation units. In this context, future work aims at exploring mathematical techniques enabling (approximate) incorporation of these complex constraints in the developed model without deteriorating significantly its computational performance.

- The ability of FD to exercise market power by revealing less time-shifting flexibility to the market is qualitatively analysed in subsection 5.4.1. It is demonstrated that such strategic behaviour has a more dominant negative impact on demand payment at the peak period than its positive impact at the off-peak period, resulting in an overall higher demand payment. As a result, consumers will find it financially unattractive to behave in such a strategic fashion. Driven by this analysis, consumers could potentially exercise market power by revealing higher time-shifting flexibility to the market. This strategic behaviour could enhance the peak demand shaving effect, thus reducing the price and demand payment at the peak period. Assuming that the demand shifting activity is energy neutral, this means that more demand will be shifted towards the off-peak period, this accordingly increases the price and demand payment. Due to the larger slope of the marginal cost of generation, the resultant payment decrement at the peak period will be more prominent than its increase at the off-peak period, leading to an overall reduction of demand payment. However, the market operator would now request the consumers to reduce their demand at the peak period according to the perceived (i.e. exaggerated) time shifting capability. This suggests that consumers' demand will be supplied at a lower quantity in the (primary) market. As a result, consumers need to arrange purchase in a *secondary market* for the shortfall of their demand. Similarly, consumers would have bought too much power at the off-peak period and consequently need to sell their excess to the secondary market. Future work will focus on devising optimal strategies for large consumers that take into account the reduced overall payment in the primary market, and the additional payment/income associated with the purchase/sale made in the secondary market.

- Chapter 4 and Chapter 5 have developed game-theoretic models to represent the strategic decision-making problems of different market players (producers, FD and ES). A future work direction would be to include uncertainties in the modelling framework. Short-term uncertainties include renewable generation, demand and competitors' actions, etc., while long-term uncertainties comprise extent and timing of generation and demand decarbonisation, cost of technologies and rivals' behaviours, regulatory policies, etc. In addition, it is necessary to take into account the fact that different market players may have different perceptions regarding the probabilities of different scenarios and follow different strategies to manage risk. To deal with these challenges, suitable stochastic and robust optimization techniques need to be incorporated in the proposed modelling framework. For instance, it has been demonstrated in Chapter 5 that a higher penetration of wind generation suppresses price levels and therefore generally limits the extent of market power exercised by strategic ES. However, the exact impact of wind generation also depends on its inherently uncertain output profile, indicating that ES operators will require suitable forecasting and stochastic optimization tools in order to devise their market strategies in future systems with high wind penetration (Section 1.1). Driven by this observation, a future extension of this work aims at incorporating such uncertainties in the game-theoretic model (Chapter 5) and employing stochastic optimization principles for its solution.

- It is also worth stressing that Chapter 4 and Chapter 5 focus on analysing the impact of participation and market power exercise of FD and ES in energy-only markets. However, in order to realize their full potential, FD and ES are envisaged to concurrently participate in markets for the provision of multiple services to several sectors in electricity industry and thus support activities related to generation, network and system operation (Section 2.4). Therefore, aggregating the value delivered by FD and ES to these sectors is vital for promoting their efficient deployment in the near future. In this context, future work aims at incorporating the provision of these system services in the presented modelling framework and investigating the interdependencies between FD and ES strategies at different market segments, under alternative market designs.

References

[1] Committee on Climate Change. Building a low-carbon economy — the UK's contribution to tackling climate change[EB/OL]. (2008-12-01)[2021-06-30]. https://www.theccc.org.uk/publication/building-a-low-carbon-economy-the-uks-contribution-to-tackling-climate-change-2/.

[2] Official Journal of the European Union. Directive 2009/28/EC of the European Parliament and of the Council of 23 April 2009 on the promotion of the use of energy from renewable sources[EB/OL]. (2015-10-05)[2021-06-30]. http://europa.eu/legislation_summaries/energy/renewable_energy/en0009_en.htm.

[3] Department of Energy and Climate Change. Digest of United Kingdom Energy Statistics 2016[EB/OL]. (2016-07-28)[2021-06-30]. https://www.gov.uk/government/collections/digest-of-uk-energy-statistics-dukes.

[4] National Grid.UK Future Energy Scenarios 2016[EB/OL]. (2016-07-28)[2021-6-30]. http://www2.nationalgrid.com/uk/industry-information/future-of-energy/future-energy-scenarios/.

[5] Department for Transport. Investigation into the Scope for the Transport Sector to Switch to Electric Vehicles and Plug-in Hybrid Vehicles[EB/OL]. (2016-10-17)[2021-06-30]. http://webarchive.nationalarchives.gov.uk/20090609003228/http://www.berr.gov.uk/files/file48653.pdf.

[6] International Energy Agency. Technology Roadmap: Electric and plug-in hybrid electric vehicles[EB/OL]. (2011-06)[2021-06-30]. http://www.iea.org/publications/freepublications/publication/technology-roadmap-electric-and-plug-in-hybrid-electric-vehicles-evphev.html.

[7] Department of Energy and Climate Change.2050 Pathways Analysis[EB/OL]. (2010-07-30)[2021-06-30]. https://www.gov.uk/guidance/2050-pathways-analysis.

[8] GAN C K, AUNEDI M, STANOJEVIC V, et al. Investigation of the impact of electrifying transport and heat sectors on the UK distribution networks[C]//21th International Conference on Electricity Distribution, Paper. 2011, 701.

[9] STRBAC G, AUNEDI M, PUDJIANTO D, et al. Strategic assessment of the role and value of energy storage systems in the UK low carbon energy future[R/OL]. (2012-06)[2021-06-30]. http://www.carbontrust.com/resources/reports/technology/energy-storage-systems-strategic-assessment-role-and-value.

[10] AUNEDI M. Value of flexible demand-side technologies in future low-carbon systems[D]. London: Imperial College London, 2013.

[11] HICKS C. Regulation of the UK electricity industry[M]. London: Centre for the Study of Regulated Industries, 1998.

[12] GREEN R J, NEWBERY D M. Competition in the British electricity spot market[J]. Journal of Political Economy, 1992, 100(5): 929-953.

[13] STRBAC G, AUNEDI M, CASTRO M, et al. Methodology for Determining Impact and Value of DSP on System Infrastructure Development[EB/OL]. (2011-09)[2021-06-30]. http://www.irene-40.eu/sites/default/files/W2IM_DV_4004_D.pdf.

[14] PAPADASKALOPOULOS D. A mechanism for decentralized participation of flexible demand in electricity markets[D]. London: Imperial College London, 2013.

[15] STRBAC G. Demand side management: Benefits and challenges[J]. Energy Policy, 2008, 36(12): 4419-4426.

[16] US Department of Energy. Benefits of demand response in electricity markets and recommendations for achieving them [R/OL]. (2006-02)[2021-06-30]. http://eetd.lbl.gov/ea/em s/reports/congress-1252d.pdf.

[17] BRADLEY P, LEACH M, TORRITI J. A review of the costs and benefits of demand response for electricity in the UK[J]. Energy Policy, 2013, 52: 312-327.

[18] PINSON P, MADSEN H. Benefits and challenges of electrical demand response: A critical review [J]. Renewable and Sustainable Energy Reviews, 2014, 39: 686-699.

[19] PUDJIANTO D, AUNEDI M, DJAPIC P, et al. Whole-systems assessment of the value of energy storage in low-carbon electricity systems[J]. IEEE Transactions on Smart Grid, 2014, 5(2): 1098-1109.

[20] ALBADI M H, EL-SAADANY E F. A summary of demand response in electricity markets[J]. Electric Power Systems Research, 2008, 78(11): 1989-1996.

[21] KIRSCHEN D S. Demand-side view of electricity markets[J]. IEEE Transactions on Power Systems, 2003, 18(2): 520-527.

[22] KHODAEI A, SHAHIDEHPOUR M, BAHRAMIRAD S. SCUC with hourly demand response considering intertemporal load characteristics[J]. IEEE Transactions on Smart Grid, 2011, 2(3): 564-571.

[23] SABER A Y, VENAYAGAMOORTHY G K. Plug-in vehicles and renewable energy sources for cost and emission reductions[J]. IEEE Transactions on Industrial Electronics, 2011, 58(4): 1229-1238.

[24] SINGH K, PADHY N P, SHARMA J. Influence of price responsive demand shifting bidding on congestion and LMP in pool-based day-ahead electricity markets[J]. IEEE Transactions on Power Systems, 2011, 26(2): 886-896.

[25] KIRSCHEN D S, STRBAC G, CUMPERAYOT P, et al. Factoring the elasticity of demand in electricity prices[J]. IEEE Transactions on Power Systems, 2000, 15(2): 612-617.

[26] SU C L, KIRSCHEN D. Quantifying the effect of demand response on electricity markets[J]. IEEE Transactions on Power Systems, 2009, 24(3): 1199-1207.

[27] WANG J, KENNEDY S, KIRTLEY J. A new wholesale bidding mechanism for enhanced demand response in smart grids[C]//2010 Innovative Smart Grid Technologies (ISGT). January 19-21, 2010, Gaither burg, MD, USA. IEEE, 2010: 1-8.

[28] LI N, HEDMAN K W. Economic assessment of energy storage in systems with high levels of renewable resources[J]. IEEE Transactions on Sustainable Energy, 2015, 6(3): 1103-1111.

[29] LAMONT A D. Assessing the economic value and optimal structure of large-scale electricity storage [J]. IEEE Transactions on Power Systems, 2013, 28(2): 911-921.

[30] SUAZO-MARTÍNEZ C, PEREIRA-BONVALLET E, PALMA-BEHNKE R, et al. Impacts of energy storage on short term operation planning under centralized spot markets [J]. IEEE Transactions on Smart Grid, 2014, 5(2): 1110-1118.

[31] BLACK M, STRBAC G. Value of bulk energy storage for managing wind power fluctuations[J]. IEEE Transactions on Energy Conversion, 2007, 22(1): 197-205.

[32] JIANG R W, WANG J H, GUAN Y P. Robust unit commitment with wind power and pumped storage hydro[J]. IEEE Transactions on Power Systems, 2012, 27(2): 800-810.

[33] SCARF H E. The allocation of resources in the presence of indivisibilities[J]. Journal of Economic Perspectives, 1994, 8(4): 111-128.

[34] RUIZ C, CONEJO A J, GABRIEL S A. Pricing non-convexities in an electricity pool[J]. IEEE Transactions on Power Systems, 2012, 27(3): 1334-1342.

[35] ARAOZ V, JÖRNSTEN K. Semi-Lagrangean approach for price discovery in markets with non-convexities[J]. European Journal of Operational Research, 2011, 214(2): 411-417.

[36] ARAOZ-CASTILLO V, JÖRNSTEN K. Alternative pricing approach for the lumpy electricity markets [C]//2010 7th International Conference on the European Energy Market. June 23-25, 2010, Madrid, Spain. IEEE, 2010: 1-6.

[37] HOGAN W W, RING B J. On minimum-uplift pricing for electricity markets[J]. Electricity Policy Group, 2003: 1-30.

[38] GRIBIK P R, HOGAN W W, POPE S L. Market-clearing electricity prices and energy uplift[EB/OL]. Cambridge, MA, 2007: 1-46.

[39] WANG G, SHANBHAG U V, ZHENG T X, et al. An extreme-point subdifferential method for convex hull pricing in energy and reserve markets — Part I: Algorithm structure [J]. IEEE Transactions on Power Systems, 2013, 28(3): 2111-2120.

[40] WANG G, SHANBHAG U V, ZHENG T X, et al. An extreme-point subdifferential method for convex hull pricing in energy and reserve markets — Part II: Convergence analysis and numerical performance[J]. IEEE Transactions on Power Systems, 2013, 28(3): 2121-2127.

[41] O'NEILL R P, SOTKIEWICZ P M, HOBBS B F, et al. Efficient market-clearing prices in markets with nonconvexities[J]. European Journal of Operational Research, 2005, 164(1): 269-285.

[42] BJØRNDAL M, JÖRNSTEN K. Equilibrium prices supported by dual price functions in markets with non-convexities[J]. European Journal of Operational Research, 2008, 190(3): 768-789.

[43] MOTTO A L, GALIANA F D. Equilibrium of auction markets with unit commitment: The need for augmented pricing[J]. IEEE Power Engineering Review, 2002, 17(3): 798-805.

[44] GALIANA F D, MOTTO A L, BOUFFARD F. Reconciling social welfare, agent profits, and consumer payments in electricity pools[J]. IEEE Transactions on Power Systems, 2003, 18(2):

452-459.

[45] BOUFFARD F, GALIANA F D. Generalized uplifts in pool-based electricity markets[M]// Analysis, control and optimization of complex dynamic systems. Springer, Boston, MA, 2005: 193-214.

[46] KIRSCHEN D, STRBAC G. Fundamentals of power system economics[M]. Chichester, UK: John Wiley & Sons, Ltd, 2004.

[47] HOBBS B F, METZLER C B, PANG J S. Strategic gaming analysis for electric power systems: An MPEC approach[J]. IEEE Transactions on Power Systems, 2000, 15(2): 638-645.

[48] WEBER J D, OVERBYE T J. An individual welfare maximization algorithm for electricity markets [J]. IEEE Transactions on Power Systems, 2002, 17(3): 590-596.

[49] GOUNTIS V P, BAKIRTZIS A G. Bidding strategies for electricity producers in a competitive electricity marketplace[J]. IEEE Transactions on Power Systems, 2004, 19(1): 356-365.

[50] WANG X, LI Y Z, ZHANG S H. Oligopolistic equilibrium analysis for electricity markets: A nonlinear complementarity approach[J]. IEEE Transactions on Power Systems, 2004, 19(3): 1348-1355.

[51] PEREIRA M V, GRANVILLE S, FAMPA M H C, et al. Strategic bidding under uncertainty: A binary expansion approach[J]. IEEE Transactions on Power Systems, 2005, 20(1): 180-188.

[52] BARROSO L A, CARNEIRO R D, GRANVILLE S, et al. Nash equilibrium in strategic bidding: A binary expansion approach[J]. IEEE Transactions on Power Systems, 2006, 21(2): 629-638.

[53] RUIZ C, CONEJO A J. Pool strategy of a producer with endogenous formation of locational marginal prices[J]. IEEE Transactions on Power Systems, 2009, 24(4): 1855-1866.

[54] POZO D, CONTRERAS J. Finding multiple Nash equilibria in pool-based markets: A stochastic EPEC approach[J]. IEEE Transactions on Power Systems, 2011, 26(3): 1744-1752.

[55] RUIZ C, CONEJO A J, SMEERS Y. Equilibria in an oligopolistic electricity pool with stepwise offer curves[J]. IEEE Transactions on Power Systems, 2012, 27(2): 752-761.

[56] European Smart Grids Technology Platform. Strategic Deployment Document for Europe's Electricity Networks of the Future[EB/OL]. (2010-04)[2021-06-30]. http://www.smartgrids.eu/documents/SmartGrids_SDD_FINAL_APRIL2010.pdf.

[57] KAZEMPOUR S J, CONEJO A J, RUIZ C. Strategic bidding for a large consumer[J]. IEEE Transactions on Power Systems, 2015, 30(2): 848-856.

[58] DARAEEPOUR A, KAZEMPOUR S J, PATIÑO-ECHEVERRI D, et al. Strategic demand-side response to wind power integration[J]. IEEE Transactions on Power Systems, 2016, 31(5): 3495-3505.

[59] SCHILL W P, KEMFERT C. Modeling strategic electricity storage: The case of pumped hydro storage in Germany[J]. The Energy Journal, 2011, 32(3): 59-87.

[60] SIOSHANSI R. Welfare impacts of electricity storage and the implications of ownership structure[J]. The Energy Journal, 2010, 31(2): 173-198.

[61] SIOSHANSI R. When energy storage reduces social welfare[J]. Energy Economics, 2014, 41: 106-

116.

[62] MOHSENIAN-RAD H. Coordinated price-maker operation of large energy storage units in nodal energy markets[J]. IEEE Transactions on Power Systems, 2016, 31(1): 786-797.

[63] KEANE A, TUOHY A, MEIBOM P, et al. Demand side resource operation on the Irish power system with high wind power penetration[J]. Energy Policy, 2011, 39(5): 2925-2934.

[64] GOMES A, ANTUNES C H, MARTINS A G. A multiple objective evolutionary approach for the design and selection of load control strategies[J]. IEEE Transactions on Power Systems, 2004, 19(2): 1173-1180.

[65] PAPADASKALOPOULOS D, STRBAC G. Decentralized participation of flexible demand in electricity markets — Part I: Market mechanism[J]. IEEE Transactions on Power Systems, 2013, 28(4): 3658-3666.

[66] PAPADASKALOPOULOS D, STRBAC G, MANCARELLA P, et al. Decentralized participation of flexible demand in electricity markets — Part II: Application with electric vehicles and heat pump systems[J]. IEEE Transactions on Power Systems, 2013, 28(4): 3667-3674.

[67] PAPADASKALOPOULOS D, STRBAC G. Decentralized optimization of flexible loads operation in electricity markets[C]//2013 IEEE Grenoble Conference. June 16-20, 2013, Grenoble, France. IEEE, 2013: 1-6.

[68] SAEZ-DE-IBARRA A, MILO A, GAZTAÑAGA H, et al. Co-optimization of storage system sizing and control strategy for intelligent photovoltaic power plants market integration[J]. IEEE Transactions on Sustainable Energy, 2016, 7(4): 1749-1761.

[69] DICORATO M, FORTE G, PISANI M, et al. Planning and operating combined wind-storage system in electricity market[J]. IEEE Transactions on Sustainable Energy, 2012, 3(2): 209-217.

[70] AWAD A S A, FULLER J D, EL-FOULY T H M, et al. Impact of energy storage systems on electricity market equilibrium[J]. IEEE Transactions on Sustainable Energy, 2014, 5(3): 875-885.

[71] YE Y J, PAPADASKALOPOULOS D, STRBAC G. An MPEC approach for analysing the impact of energy storage in imperfect electricity markets[C]//2016 13th International Conference on the European Energy Market (EEM) June 6-9, 2016, Porto, Portugal. IEEE, 2016: 1-5.

[72] MAS-COLELL A, WHINSTON M D, GREEN J R. Microeconomic theory[M]. New York: Oxford university press, 1995.

[73] DE JONGHE C, HOBBS B F, BELMANS R. Optimal generation mix with short-term demand response and wind penetration[J]. IEEE Transactions on Power Systems, 2012, 27(2): 830-839.

[74] SIOSHANSI R. Evaluating the impacts of real-time pricing on the cost and value of wind generation[J]. IEEE Transactions on Power Systems, 2010, 25(2): 741-748.

[75] PARVANIA M, FOTUHI-FIRUZABAD M, SHAHIDEHPOUR M. Optimal demand response aggregation in wholesale electricity markets[J]. IEEE Transactions on Smart Grid, 2013, 4(4): 1957-1965.

[76] GKATZIKIS L, KOUTSOPOULOS I, SALONIDIS T. The role of aggregators in smart grid demand response markets[J]. IEEE Journal on Selected Areas in Communications, 2013, 31(7):

1247-1257.

[77] TAN Xa, LI Q M, WANG H. Advances and trends of energy storage technology in microgrid[J]. International Journal of Electrical Power & Energy Systems, 2013, 44(1): 179-191.

[78] CONEJO A J, MORALES J M, BARINGO L. Real-time demand response model[J]. IEEE Transactions on Smart Grid, 2010, 1(3): 236-242.

[79] O'CONNELL N, WU Q W, ØSTERGAARD J, et al. Day-ahead tariffs for the alleviation of distribution grid congestion from electric vehicles[J]. Electric Power Systems Research, 2012, 92: 106-114.

[80] DICORATO M, FORTE G, PISANI M, et al. Planning and operating combined wind-storage system in electricity market[J]. IEEE Transactions on Sustainable Energy, 2012, 3(2): 209-217.

[81] AKHAVAN-HEJAZI H, MOHSENIAN-RAD H. Optimal operation of independent storage systems in energy and reserve markets with high wind penetration[J]. IEEE Transactions on Smart Grid, 2014, 5(2): 1088-1097.

[82] PAPADASKALOPOULOS D, PUDJIANTO D, STRBAC G. Decentralized coordination of microgrids with flexible demand and energy storage[J]. IEEE Transactions on Sustainable Energy, 2014, 5(4): 1406-1414.

[83] PAPADASKALOPOULOS D, STRBAC G. Nonlinear and randomized pricing for distributed management of flexible loads[J]. IEEE Transactions on Smart Grid, 2016, 7(2): 1137-1146.

[84] DRURY E, DENHOLM P, SIOSHANSI R. The value of compressed air energy storage in energy and reserve markets[J]. Energy, 2011, 36(8): 4959-4973.

[85] KARANGELOS E, BOUFFARD F. Towards full integration of demand-side resources in joint forward energy/reserve electricity markets[J]. IEEE Transactions on Power Systems, 2012, 27(1): 280-289.

[86] MORENO R, MOREIRA R, STRBAC G. A MILP model for optimising multi-service portfolios of distributed energy storage[J]. Applied Energy, 2015, 137: 554-566.

[87] PEREZ A, MORENO R, MOREIRA R, et al. Effect of battery degradation on multi-service portfolios of energy storage[J]. IEEE Transactions on Sustainable Energy, 2016, 7(4): 1718-1729.

[88] HAN S, HAN S, SEZAKI K. Development of an optimal vehicle-to-grid aggregator for frequency regulation[J]. IEEE Transactions on Smart Grid, 2010, 1(1): 65-72.

[89] HARVEY S M, HOGAN W W. Market power and withholding[J]. Harvard Univ., Cambridge, MA, 2001.

[90] CARRIÓN M, ARROYO J M. A computationally efficient mixed-integer linear formulation for the thermal unit commitment problem[J]. IEEE Transactions on Power Systems, 2006, 21(3): 1371-1378.

[91] YE Y J, PAPADASKALOPOULOS D, STRBAC G. Factoring flexible demand non-convexities in electricity markets[J]. IEEE Transactions on Power Systems, 2015, 30(4): 2090-2099.

[92] CHIANG J, LEE L F. Discrete/continuous models of consumer demand with binding nonnegativity constraints[J]. Journal of Econometrics, 1992, 54(1/2/3): 79-93.

[93] ZHUANG F, GALIANA F D. Towards a more rigorous and practical unit commitment by Lagrangian relaxation[J]. IEEE Transactions on Power Systems, 1988, 3(2): 763-773.

[94] DEKRAJANGPETCH S, SHEBLE G B, CONEJO A. Auction implementation problems using Lagrangian relaxation[J]. IEEE Transactions on Power Systems, 1999, 14(1): 82-88.

[95] ONGSAKUL W, PETCHARAKS N. Unit commitment by enhanced adaptive Lagrangian relaxation [J]. IEEE Transactions on Power Systems, 2004, 19(1): 620-628.

[96] REDONDO N J, CONEJO A J. Short-term hydro-thermal coordination by Lagrangian relaxation: solution of the dual problem[J]. IEEE Transactions on Power Systems, 1999, 14(1): 89-95.

[97] MADRIGAL M, QUINTANA V H. An interior-point/cutting-plane method to solve unit commitment problems[J]. IEEE Transactions on Power Systems, 2000, 15(3): 1022-1027.

[98] BORGHETTI A, FRANGIONI A, LACALANDRA F, et al. Lagrangian heuristics based on disaggregated Bundle methods for hydrothermal unit commitment[J]. IEEE Transactions on Power Systems, 2003, 18(1): 313-323.

[99] BAZARAA M S, SHERALI H D, SHETTY C M. Nonlinear programming: theory and algorithms [M]. Hoboken, NJ, USA: John Wiley & Sons, Inc., 2005.

[100] TWOMEY P, GREEN R, NEUHOFF K, et al. A review of the monitoring of market power the possible roles of tsos in monitoring for market power issues in congested transmission systems, [J]. Cambridge Working Papers in Economics, 2005.

[101] STOFT S. Power system economics: designing markets for power[M]. Piscataway, NJ: IEEE press, 2002.

[102] DAVID A K, WEN F S. Market power in electricity supply[J]. IEEE Transactions on Energy Conversion, 2001, 16(4): 352-360.

[103] BORENSTEIN S, BUSHNELL J B, WOLAK F A. Measuring market inefficiencies in California's restructured wholesale electricity market[J]. American Economic Review, 2002, 92(5): 1376-1405.

[104] LI G, SHI J, QU X L. Modeling methods for GenCo bidding strategy optimization in the liberalized electricity spot market: A state-of-the-art review[J]. Energy, 2011, 36(8): 4686-4700.

[105] KLEMPERER P D, MEYER M A. Supply function equilibria in oligopoly under uncertainty[J]. Econometrica: Journal of the Econometric Society, 1989, 57(6): 1243-1277.

[106] BALDICK R. Electricity market equilibrium models: The effect of parameterization[J]. IEEE Power Engineering Review, 2002, 22(7): 53.

[107] YOUNES Z, ILIC M. Generation strategies for gaming transmission constraints: Will the deregulated electric power market be an oligopoly? [J]. Decision Support Systems, 1999, 24(3/4): 207-222.

[108] GABRIEL S A, CONEJO A J, FULLER J D, et al. Complementarity modeling in energy markets [M]. Springer Science & Business Media, 2012.

[109] KAZEMPOUR S J. Strategic generation investment and equilibria in oligopolistic electricity markets [D]. Spain: Department of Electric, Electronic, Automatic and Communication

Engineering, Universidad de Castilla-La Mancha, 2013.

[110] HARRIS C. Electricity markets: pricing, structures and economics[M]. Oxford, UK: John Wiley & Sons, 2006.

[111] BIGGAR D R, HESAMZADEH M R. The economics of electricity markets[M]. Chichester, UK: John Wiley & Sons Ltd., 2014.

[112] NEWBERY D. Mitigating market power in electricity networks[J]. Department of Applied Economics, University of Cambridge, mimeo, 2002.

[113] NANDURI V, DAS T K. A survey of critical research areas in the energy segment of restructured electric power markets[J]. International Journal of Electrical Power & Energy Systems, 2009, 31 (5): 181-191.

[114] WOLAK F A. An empirical analysis of the impact of hedge contracts on bidding behavior in a competitive electricity market[J]. International Economic Journal, 2000, 14(2): 1-39.

[115] BERRY G A, HOBBS B F, MERONEY W A, et al. Analysing strategic bidding behaviour in transmission networks[J]. Util. Pol., 1999, 8(3): 139 - 158.

[116] ACKERMANN T. Distributed resources and re-regulated electricity markets[J]. Electric Power Systems Research, 2007, 77(9): 1148-1159.

[117] BOMPARD E, MA Y C, NAPOLI R, et al. The demand elasticity impacts on the strategic bidding behavior of the electricity producers[J]. IEEE Transactions on Power Systems, 2007, 22(1): 188-197.

[118] BOMPARD E, MA Y C, NAPOLI R, et al. The impacts of price responsiveness on strategic equilibrium in competitive electricity markets[J]. International Journal of Electrical Power & Energy Systems, 2007, 29(5): 397-407.

[119] BOMPARD E, NAPOLI R, WAN B. The effect of the programs for demand response incentives in competitive electricity markets[J]. European Transactions on Electrical Power, 2009, 19(1): 127-139.

[120] THIMMAPURAM P R, KIM J, BOTTERUD A, et al. Modeling and simulation of price elasticity of demand using an agent-based model[C]//2010 Innovative Smart Grid Technologies (ISGT). January 19-21, 2010, Gaithersburg, MD, USA. IEEE, 2010: 1-8.

[121] London Economics. The Value of Lost Load (VoLL) for Electricity in Great Britain: Final report for OFGEM and DECC[R/OL]. (2013-07)[2021-06-30]. https://www.ofgem.gov.uk/ofgem-publications/82293/london-economics-value-lost-load-electricity-gbpdf.

[122] FORTUNY-AMAT J, MCCARL B. A representation and economic interpretation of a two-level programming problem[J]. Journal of the Operational Research Society, 1981, 32(9): 783-792.

[123] FAN Y, PAPADASKALOPOULOS D, STRBAC G. A game theoretic modelling framework for decentralized transmission planning[C]//2016 Power Systems Computation Conference (PSCC). June 20-24, 2016, Genoa, Italy. IEEE, 2016: 1-7.

[124] SIOSHANSI R, DENHOLM P, JENKIN T. Market and policy barriers to deployment of energy storage[J]. Economics of Energy & Environmental Policy, 2012, 1(2): 47-64.

[125] STRBAC G, RAMSAY C, PUDJIANTO D. Framework for development of enduring UK transmission access arrangements[R/OL]. (2007-07)[2021-06-30]. https://www.scribd.com/document/135837099/Framework-for-development-of-enduring-UK-transmission-access-arrangements-pdf.

[126] National Grid. 2010 Seven Year Statement[EB/OL]. (2010)[2021-06-30]. http://nationalgrid.com/.

[127] FICO XPRESS website[EB/OL]. [2021-06-30]. http://www.fico.com/en/Products/DMTools/Pages/FICO-Xpress-Optimization-Suite.aspx.

[128] BOYD S, VANDENBERGHE L. Convex optimization[M]. Cambridge: Cambridge university press, 2004.

[129] NOCEDAL J, WRIGHT S. Numerical optimization[M]. Springer Science & Business Media, 2006.

[130] BAZARAA M S, SHERALI H D, SHETTY C M. Nonlinear programming: theory and algorithms [M]. John Wiley & Sons, 1993.

Appendix A
Convexity Principles

The term "*convexity*" generally applies to sets, functions and optimization problems[128-129].

A set $C \in \mathbb{R}^n$ is a *convex* if the line segment between any two points in C lies in C, i.e. if for any x_1, $x_2 \in C$ and any θ with $0 \leqslant \theta \leqslant 1$, we have:

$$\theta x_1 + (1-\theta)x_2 \in C \tag{A.1}$$

The *convex hull* of a set C, denoted **conv** C, is the set of all convex combinations of points in C:

$$\mathbf{conv} = \{\theta_1 x_1 + \cdots + \theta_k x_k \mid x_i \in C, \theta_i \geqslant 0, i=1, \cdots, k, \theta_1 + \cdots + \theta_k = 1\} \tag{A.2}$$

As the name suggests, the convex hull **conv** C is always convex. It is the smallest convex set that contains C: if B is any convex set that contains C, then **conv** $C \subseteq B$.

A function $f: \mathbb{R}^n \to \mathbb{R}$ is a convex if **dom** f is a convex set and if for all x, $y \in$ **dom** f, and θ with $0 \leqslant \theta \leqslant 1$, we have:

$$f(\theta x + (1-\theta)y) \leqslant \theta f(x) + (1-\theta)f(y) \tag{A.3}$$

A function f is *strictly convex* if strict inequality holds in (A.3) whenever $x \neq y$ and $0 < \theta < 1$. We say f is concave if $-f$ is convex, and *strictly concave* if $-f$ is strictly convex.

A *convex optimization problem* is one of the form:

$$\min f_0(x) \tag{A.4}$$

$$\text{s.t.} \ f_i(x) \leqslant b_i, i=1, \cdots, m \tag{A.5}$$

where the functions $f_0, \cdots, f_m: \mathbb{R}^n \to \mathbb{R}$ are convex, i.e. satisfy

$$f_i(\alpha x + \beta y) \leqslant \alpha f_i(x) + \beta f_i(y) \tag{A.6}$$

for all x, $y \in \mathbb{R}^n$ and all α, $\beta \in \mathbb{R}$ with $\alpha + \beta = 1$, $\alpha \geqslant 0$, $\beta \geqslant 0$.

An optimization problem is strictly convex if its feasible domain is convex and its objective function is strictly convex.

Appendix B
Lagrangian Formulation of Convex Hull Pricing Problem

Consider a generic optimization problem for a unit comment economic dispatch problem:

$$\min_{x \in X} f(x) \quad (B.1)$$

$$\text{s.t. } g(x) = y \quad (B.2)$$

The load vector y must be met by the choice of decision variable x that comprises of the generation dispatches and unit commitment statuses. The constraint function $g(x)$ describes interactions across many separate unit decisions, while the feasible set X characterizes the individual operational constraints of the generating units (including binary constraints). The objective function $f(x)$ represents the cost of commitment and dispatch to meet load.

As load varies, the value of the least-cost solution changes accordingly. Define the *value function* (i.e. minimum cost function) as:

$$v(y) = \min_{x \in X}\{f(x) \mid g(x) = y\} \quad (B.3)$$

The value function plays a central role in the definition and derivation of equilibrium prices. The slope of the value function represents the marginal cost of meeting an additional unit of load. Along with the more general derivation when both load and generation are optimized, this marginal cost defines the market-clearing price.

Let us introduce the vector of price λ and the Lagrangian function associated with the primal optimization problem (B.1)—(B.2):

$$L(y, x, \lambda) = f(x) + \lambda(y - g(x)) \quad (B.4)$$

The Lagrangian factors out the complicating constraints and, given λ, produces a problem that is easier to solve. For given prices, define the optimized Lagrangian value as:

$$\hat{L}(y, \lambda) = \min_{x \in X} L(y, x, \lambda) = \min_{x \in X}\{f(x) + \lambda(y - g(x))\} \quad (B.5)$$

The associated dual problem is defined as choosing the price λ^D to maximize the optimized Lagrangian to obtain:

$$L^*(y) = \max_\lambda \widehat{L}(y, \lambda) = \max_\lambda \{\min_{x \in X}\{f(x) + \lambda(y - g(x))\}\} \qquad (B.6)$$

In the case of a well-behaved convex optimization problem, where the decision variables are continuous and the constraint sets are all convex, the optimal dual solution produces a vector of prices that supports the optimal solution. In particular, using these prices, the corresponding solution for x embedded in $v(y)$ also solves the problem in $\widehat{L}(y, \lambda)$. Furthermore, under these conditions, we have:

$$v(y) = L^*(y) \qquad (B.7)$$

In the more general situation (without the convenient convexity assumptions), there may be no equilibrium prices that support the solution at y and results in a duality gap, where:

$$v(y) > L^*(y) \qquad (B.8)$$

To make the connection with the minimum uplift, consider this formulation of the dual problem. By definition, for the dual solution we have:

$$L^*(y) = \max_\lambda \{\min_{x \in X}\{f(x) + \lambda(y - g(x))\}\}$$
$$\leq \max_\lambda \{\min_{x \in X}\{f(x) + \lambda(y - g(x)) \mid g(x) = y\}\}$$
$$= \max_\lambda \{\min_{x \in X}\{f(x) \mid g(x) = y\}\} = \min_{x \in X}\{f(x) \mid g(x) = y\} = v(y)$$
$$\qquad (B.9)$$

The difference $v(y) - L^*(y)$ is known as the duality gap. When equality holds, as when $v(y)$ is convex and certain regularity conditions hold, there is no duality gap[129].

Let us now consider the following alternative form of the optimized Langrangian:

$$\widehat{L}(y, \lambda) = \min_{x \in X}\{f(x) + \lambda(y - g(x))\}$$
$$= \min_{x \in X, z}\{f(x) + \lambda(y - z) \mid g(x) = z\}$$
$$= \lambda y + \min_{x \in X, z}\{f(x) - pz \mid g(x) = z\} \qquad (B.10)$$
$$= \lambda y + \min_z\{-\lambda z + \min_{x \in X, z}\{f(x) \mid g(x) = z\}\}$$
$$= \lambda y + \min_z\{-\lambda z + v(z)\}$$
$$= \lambda y - \max_z\{\lambda z - v(z)\}$$

The Fenchel convex conjugate of v is by definition[130]:

$$v^c(\lambda) = \max_z \{\lambda z - v(z)\} \tag{B.11}$$

Note that $v^c(\lambda)$ is the supremum over a set of convex (affine) functions of λ and is therefore a convex function of λ. Hence, we have:

$$\widehat{L}(y, \lambda) = \lambda y - v^c(\lambda) \tag{B.12}$$

Now

$$L^*(y) = \max_\lambda \widehat{L}(y, \lambda) = \max_\lambda \{\lambda y - v^c(\lambda)\} \tag{B.13}$$

Therefore, applying the conjugate definition again, we have:

$$L^*(y) = v^{cc}(y) \tag{B.14}$$

The resulting function $v^{cc}(y)$ is a closed convex function of y and we have:

$$v^{cc}(y) = L^*(y) \leqslant v(y) \tag{B.15}$$

Suppose that $\overline{v}(y)$ is the closed convex hull of $v(y)$. Then we have:

$$\overline{v}(y) = v^{cc}(y) \leqslant v(y) \tag{B.16}$$

In the other words, $v^{cc}(\lambda)$ equals the convex hull of $v(y)$ over y. Futhermore, under one of the regularity conditions with λ^D as a solution to the dual problem, λ^D defines a supporting hyperplane[128], [130] (i.e. the marginal cost) for $v^{cc}(y)$ at y:

$$v^{cc}(\lambda) = \max_\lambda \{\lambda y - v^c(\lambda)\}$$

$$\geqslant \lambda^D z - v^c(\lambda^D) \tag{B.17}$$
$$= \lambda^D z + v^{cc}(y) - \lambda^D y$$
$$= v^{cc}(y) + \lambda^D(z - y)$$

Therefore, λ^D constitutes a sub-gradient of $v^{cc}(y) = \overline{v}(y)$, that is:

$$\lambda^D \in \partial \overline{v}(y) \tag{B.18}$$

In general, the price supports defined by the sub-gradients are not unique, but all elements of the set characterized by the subdifferential, $\partial \overline{v}(y)$, support the convex hull.

The connection to the uplift depends on a certain economic interpretation of the duality gap. From above:

$$L^*(y) = \max_\lambda \widehat{L}(y, \lambda) = \max_\lambda \{\lambda y - \max_z \{\lambda z - v(z)\}\} \tag{B.19}$$

Therefore, the duality gap DG can be expressed as:

$$\begin{aligned} DG &= v(y) - L^*(y) \\ &= v(y) - \max_\lambda \{\lambda y - \max_z \{\lambda z - v(z)\}\} \\ &= v(y) + \min_\lambda \{-\lambda y + \max_z \{\lambda z - v(z)\}\} \\ &= \min_\lambda \{\max_z \{\lambda z - v(z)\} - [\lambda y - v(y)]\} \end{aligned} \quad (B.20)$$

Given any output z, the economic profit is:

$$pro(\lambda, z) = \lambda z - v(z) \quad (B.21)$$

This is the difference between the revenues for quantity z at price λ and the minimum cost of meeting the requirement in z. We give an economic interpretation where the first term in the duality gap is the profit maximizing outcome given price λ:

$$pro^*(\lambda) = \max_z \{\lambda z - v(z)\} = \max_z pro(\lambda, z) \quad (B.22)$$

The actual economic profit without further uplift payments is expressed as:

$$pro(\lambda, y) = \lambda y - v(y) \quad (B.23)$$

If we have to make up the difference in order to compensate direct losses or for foregone opportunities, then the total payment is the difference:

$$Uplift(\lambda, y) = pro^*(\lambda) - pro(\lambda, y) \quad (B.24)$$

In other words, the dual problem seeks price λ^D that minimizes the uplift (B.24). Under this interpretation, the duality gap equals the minimum uplift across all possible prices λ:

$$\begin{aligned} v(y) - L^*(y) &= \min_\lambda \{\max_z \{\lambda z - v(z)\} - [\lambda y - v(y)]\} \\ &= \min_\lambda \{pro^*(\lambda) - pro(\lambda, y)\} \\ &= \min_\lambda Uplift(\lambda, y) \end{aligned} \quad (B.25)$$

Therefore, if λ^D is a dual solution and

$$\lambda^D \in \arg\max_z \{\lambda^D z - v(z)\} \quad (B.26)$$

then there is no duality gap and thus no uplift. In this sense, the price λ^D support the competitive equilibrium solution if there is no duality gap. However, if

$$\lambda^D \notin \arg\max_z \{\lambda^D z - v(z)\} \quad (B.27)$$

then λ^D does not support the equilibrium solution (i.e. $\arg\max_z\{\lambda^D z - v(z)\} \neq y$), and there exists a duality gap equal to the minimum uplift. It should be noted that the uplift under the definition (B.24) is always nonnegative. Uniform linear prices differ themselves regarding the amount of the requisite uplift payments to support the competitive equilibrium. In this regard, the convex hull prices are corresponding to the minimum uplift payment.

Appendix C
List of Abbreviations

CAES	Compressed Air Energy Storage
CCFD	Continuous Controllable Flexible Demand
CCGT	Combined Cycle Gas Turbine
CCS	Carbon Capture Storage
CPP	Critical Peak Pricing
DCOPF	Direct Current Optimal Power Flow
DUDI	Demand Utility Deviation Index
ED	Economic Dispatch
EHP	Electrical Heat Pump
EPEC	Equilibrium Program with Equilibrium Constraints
ES	Energy Storage
EV	Electric Vehicle
FCFD	Fixed Cycle Flexible Demand
FD	Flexible Demand
GHG	Green House Gas
GPDI	Generation Profit Deviation Index
GTM	Game Theoretic Model
ICT	Information and Communication Technologies
KKT	Karush-Kuhn-Tucker
LMP	Locational Marginal Price
LR	Lagrangian Relaxation
MCP	Market Clearing Price
MII	Market Inefficiency Index
MILP	Mixed-Integer Linear Programming
MPEC	Mathematical Program with Equilibrium Constraints
MSG	Minimum Stable Generation
NE	Nash Equilibrium
PEV	Plug-in Electric Vehicle
RES	Renewable Energy Source

RTP	Real Time Pricing
SCUC	Security Constrained Unit Commitment
SFE	Supply Function Equilibrium
SLR	Semi Lagrangian Relaxation
SME	Small & Medium Enterprises
SOC	State Of Charge
SPDI	Storage Profit Deviation Index
T&D	Transmission and Distribution
TOU	Time Of Use
V2G	Vehicle to Grid
VoLL	Value of Lost Load
WA	Wet Appliance